The Structure of
Lebesgue Integration Theory

The Structure of Lebesgue Integration Theory

G. TEMPLE

OXFORD
AT THE CLARENDON PRESS
1971

Oxford University Press, Ely House, London W. 1

GLASGOW NEW YORK TORONTO MELBOURNE WELLINGTON
CAPE TOWN SALISBURY IBADAN NAIROBI DAR ES SALAAM LUSAKA ADDIS ABABA
BOMBAY CALCUTTA MADRAS KARACHI LAHORE DACCA
KUALA LUMPUR SINGAPORE HONG KONG TOKYO

PRINTED IN GREAT BRITAIN

Preface

The purpose of this work is to introduce the principles and techniques of the theory of integration in the general and simple form that we owe primarily to Lebesgue, de la Vallée-Poussin, and W. H. Young. It is addressed to those who are already familiar with the elementary calculus of differentiation and integration as applied to the standard functions of algebraical and trigonometrical type. Some slight acquaintance with the topology of open and closed sets may also now be presumed in most first-year undergraduates, for whom the book is written, but it is not essential. I have endeavoured to provide an account of the essentials of the theory and practice of Lebesgue integration that are indispensable in analysis, in theoretical physics, and in the theory of probability in a form that can be readily assimilated by students reading for honours in mathematics, physics, or engineering.

To realize this purpose is a serious and important pedagogical problem, for the theory of Lebesgue integration occupies a strange, ambivalent position in the minds of mathematicians confronted with the challenge of planning a syllabus for undergraduates. Then Lebesgue integration appears to be at once indispensable and unattainable, desirable and impracticable.

As compared with 'Riemann' integration, so strongly entrenched in university courses of analysis, the subject of 'Lebesgue' integration possesses three great advantages: it is applicable to a much larger class of functions, the properties of the integral are much easier to establish, and the applications of the theory are made with much greater facility. And yet the Lebesgue theory is almost universally regarded as too difficult for inclusion in undergraduate instruction, and in spite of the numerous excellent expositions of the Lebesgue theory, there is still a need for a strictly elementary account of the subject, which will make it readily accessible and utilizable in an undergraduate course.

First of all let us squarely face the ineluctable problems which confront the writer who aspires to provide what the French so happily call 'une œuvre de la haute vulgarisation'. It must be admitted that there are undoubtedly two arduous passages in the traditional approach to the Lebesgue theory. The first is the theory of measure and the second is the theory of the differentiation of an indefinite integral.

The first difficulty can be evaded by framing a direct definition of the integral, independently of the theory of measure. This has been done by L. C. Young, by O. Perron, and by F. Riesz and B. Sz-Nagy. For a number of cogent reasons I have resisted the temptation to follow this seductive deviation from the traditional route.

In the first place in the systematic and structural account of the Lebesgue theory the theory of measure is conceptually prior to the general theory of integration, since it is in fact the theory of the integration of the simple functions whose range consists of just two numbers, zero and unity. In the second place it is impossible to avoid the concept of sets of points of 'zero measure'. In the third place the theory of measure is indispensable in such important applications as ergodic theory and the theory of probability.

In the present introduction to integration theory the theory of measure has therefore been retained, but an attempt has been made to simplify and shorten the exposition by translating the traditional account from geometrical into analytical language. This is easily accomplished by systematically representing a set of points E by its characteristic function $\chi(E)$—a device suggested by Ch. J. de la Vallée-Poussin.

The theory has also been simplified by replacing 'open sets' as the central concept, by enumerable collections of intervals which may be closed, open, or half-open.

The second serious difficulty in the Lebesgue theory is the differentiation of an indefinite integral. In elementary calculus an integral $\phi(x)$ of a function $f(x)$ is descriptively defined as a function whose derivative is the integrand $f(x)$. To students familiar with this concept it must be a sad and disheartening experience to realize that the corresponding property of the Lebesgue integral is almost the last to be established, and that it requires the formidable apparatus of the covering theorems of Vitali or of Riesz, or the theory of *réseaux* developed by de la Vallée-Poussin.

In the method of treatment proposed in this book the theory of differentiation is based on the analytical discussion given by F. Riesz and B. Sz-Nagy (1953) and its geometrical expression in the 'rising sun' theorem of F. Boas (1960).

The removal of these two well-known difficulties in the Lebesgue theory is scarcely an adequate excuse for the publication of yet another introduction to this much-introduced subject. The real justification lies in the more exacting demands now made on authors of mathematical works. The revolution in mathematical theory, which is still proceeding, has also provoked a revolution in mathematical teaching, and has imposed new canons of exposition. The two essential and necessary conditions which a modern textbook must attempt to satisfy are those associated with the key-words 'motivation' and 'structure'.

An exposition of any branch of mathematics must now provide the student with an adequate motivation, that is with a line of thought which leads naturally, and almost inevitably and automatically, from the elementary concepts and methods he already possesses to the more general and abstract ideas and techniques of the theory that he proposes to study. The motivation reveals the inadequacy of our present knowledge, it poses urgent and important questions we are as yet unable to solve, and, in its highest achievements, it restates these questions in a form which suggests what methods must be devised for a solution.

The appetites excited by motivation must then be satisfied by a systematic exposition in which the whole of the theory is dominated by a few simple principles that endow the subject with a definite structure that can be described in advance before the student is committed to a detailed study. Thus a structure is not so much a set of definitions and theorems as a programme that directs the advance of the whole subject.

These are high ideals for any writer and the present book must be regarded as an experiment designed to determine how far these ideals can be realized in an introduction to the subject of integration. This is a subject which now offers most appropriate material for such an investigation. Numerous accounts of the theory of integration have been published, each of them furnishing its own special insight and technique. It now seems possible to extract the essential motivating ideas and structural principles that unify the whole theory.

The motivating ideas, described in Chapter 2, lead from the Archimedean 'method of exhaustion' to the general concept of an integral.

The structural principles are two in number—here termed the principle of bracketing and the principle of monotony.

The principle of bracketing is a method of induction which enables us to extend a class of functions which are susceptible of integration, by using the concept of upper and lower integrals. We can thus ascend from the concept of area or volume to the Lebesgue measure and the Lebesgue integral.

To carry out this programme we need the principle of monotony by which all sequences are reduced to monotone sequences and all functions are reduced to monotone functions. The problems of convergence and of integration are thus reduced to their simplest possible form.

Both of these principles are firmly embedded in the literature, especially in the writings of W. H. Young and L. C. Young. The main contribution of the present work is the exhibition of these concepts as the essential structure of the Lebesgue theory of measure, of integration, of differentiation, and of convergence.

I must express my indebtedness to friends and colleagues who have read this book in proof or in typescript, and who have given me valuable advice and help. I will mention especially Dr. J. D. M. Wright, who read the book in proof, and Dr. A. Ingleton, who, with great patience and much friendly criticism, has given me invaluable help in correcting and improving the successive drafts of the work.

Oxford   G. T.
20 January 1971

Contents

1 Motivation

1.1. Introduction

The study of any branch of mathematics is essentially a guided
research, and, before he commits himself to such a project, the
prudent student will require satisfactory answers to three or
four questions:

(i) What is the object of the investigation?

(ii) What measure of success will be attained?

(iii) What methods of investigation will be needed? and

(iv) What other lines of investigation are available?

We proceed to answer these questions so far as they relate to
the study of the Lebesgue theory of integration and differentiation.

1.2. The object of the Lebesgue theory

The object of the Lebesgue theory is stated clearly by Henri
Lebesgue in the paper published in 1902 which is his thesis—
certainly the most famous and influential doctoral thesis ever
written. There he states his purpose to give the most precise and
general definitions of three mathematical concepts: the integral
of a function, the length of a curve, and the area of a curved
surface.

But why seek for the most general definition of an integral?
Why not be content with the integral as defined in elementary
treatises on the calculus?

There are two reasons for a divine discontent.

(i) The greater generality of a definition is obtained by
greater abstraction and therefore with greater simplicity.

(ii) The familiar world of the well-behaved 'tame' functions
of elementary calculus is not 'closed', and most limiting processes
take us out of these comfortable surroundings into a strange

world of 'wild' functions where the elementary concepts of integration are no longer valid.

In view of the notorious difficulty of the Lebesgue theory the first reason may appear an idle paradox. But for the purpose of comprehending the real nature of an integral we know far too much about the properties of particular functions such as x^n, $\sin x$ and $\cos x$, $\exp x$ and $\log x$, and the multitude of known facts is an embarrassment. As soon as we begin to generalize and abstract we are no longer concerned with these trivial details and we can concentrate on the essential features of the problem, for example, is the function to be integrated monotone or continuous? The general definition of an integral given by Lebesgue is in fact essentially simple because it depends only on the most general properties of the function to be integrated. (In fairness to the student it must, however, be admitted that simplicity in mathematics, like simplicity of character, is an ideal to be achieved only by unremitting toil.)

The construction of 'wild' functions from 'tame' functions by means of limiting processes is now an integral part of analysis with its own recognizable techniques such as the 'principle of the condensation of singularities'. Thus, from the function

$$f(x) = x \cos \ln|x| \quad (x \neq 0)$$

or $0 \qquad\qquad (x = 0),$

which has no derivative at the origin $(x = 0)$, we can construct the function

$$\phi(x) = \sum_{n=1}^{\infty} \frac{f(x-z_n)}{n^2},$$

where $z_1, z_2,...$ are the rational numbers between 0 and 1. The function $\phi(x)$ is continuous but has no derivative at any of the infinite number of points $x = z_n$, $n = 1, 2,...$.

The contemplation of such wild or pathological functions was repugnant to many classical analysts, such as Poincaré and Hermite (Saks 1937, p. iv) but their construction has two definite advantages. At the beginning of our studies it demonstrates the inadequacy of elementary analysis, and at the end of our studies it can show that the results obtained are the 'best possible'.

1.3. The achievement of Lebesgue

In this book we are concerned mainly with Lebesgue's attempt to provide the most general definition of the integral of a function of one or more variables, x, or $x_1, x_2, ..., x_n$.

First let us consider the problem of integration of *bounded* functions over a *bounded* interval. In this case the method of Lebesgue yields the most general possible definition, i.e. it applies to the widest possible class of functions. These are the functions described by Lebesgue as 'measurable'. The scope of Lebesgue's definition of the integral

$$\int_a^b f(x)\, \mathrm{d}x$$

is established by the proof that if $f(x)$ is bounded and measurable, then the indefinite Lebesgue integral

$$\phi(x) = \int_a^x f(t)\, \mathrm{d}t$$

possesses 'almost everywhere' a derivative $\phi'(x)$ equal to $f(x)$.

Without anticipating the precise definition of the phrase 'almost everywhere', it is sufficient to state that it allows $\phi(x)$ to have no derivative at a set of points that may be enumerable, as in the case of the wild function quoted above, or even non-enumerable.

Thus the object of Lebesgue's theory is completely attained so far as bounded functions over a bounded interval are concerned.

In the case of unbounded functions or unbounded intervals the success of the Lebesgue theory is only partial. In fact the Lebesgue theory now applies directly only to the limited class of functions described as 'summable', and does not apply to the class of integrals which are not 'absolutely convergent'. Thus the integral

$$\int_0^\infty \left| \frac{\sin x}{x} \right| \mathrm{d}x$$

is infinite, and in consequence, the integral

$$\int_0^\infty \frac{\sin x}{x}\, dx$$

is not directly integrable by the methods of Lebesgue, although it may be defined by the limiting process

$$\lim_{a\to\infty} \int_0^a \frac{\sin x}{x}\, dx,$$

which is not, strictly speaking, part of the Lebesgue theory.

Finally, we must emphasize that the Lebesgue theory applies to functions of several variables and to their integrals, not only over 'domains' but also over sets of points that belong to the class described as 'measurable'.

1.4. The techniques of Lebesgue theory

The achievements of the Lebesgue theory are finely described in all the standard works on the subject, but it is not always made clear that this success of the theory is attained by the skilful use of two simple techniques, which may be described as the 'method of bracketing' and the 'method of monotony'. These methods will be expounded in detail later but they may be summarily described as follows:

The method of monotony consists in the reduction of problems of convergence to the study of monotone sequences, which are familiar and simple instruments of calculation.

The method of bracketing consists in using the integrals of certain tame functions to integrate certain wild functions by 'bracketing' a wild function between a pair of tame functions. Thus two integrable tame functions $\lambda(x)$ and $\mu(x)$ may be said to bracket a wild function $f(x)$ with tolerance $\epsilon > 0$, if

$$\lambda(x) \leqslant f(x) \leqslant \mu(x)$$

and if
$$\int_a^b \mu(x)\, dx - \int_a^b \lambda(x)\, dx < \epsilon.$$

The integrals of $\lambda(x)$ and $\mu(x)$ may then be regarded as approximations to the (as yet) unknown and undefined integral of $f(x)$.

If there exists such a pair of integrable bracketing functions for *any* prescribed tolerance $\epsilon > 0$, we have at our disposal the method to define and to evaluate the integral of $f(x)$ to any desired degree of approximation. These two techniques are all that is necessary to construct the whole of the Lebesgue theory of integration.

1.5. Alternative theories

The account of the Lebesgue integral given in the following chapters follows the mature thought of Lebesgue as expounded in the second edition (1928) of his book modestly entitled *Leçons sur l'intégration*. In this account the theory of measure is developed as a preliminary to the theory of integration, but there are alternative methods of which the student should be informed.

Broadly speaking, these alternative methods fall into two classes—in one class integration is reduced to measure theory, and, in the other, measure theory is subsumed into the theory of integration.

Lebesgue himself adopted the first method in the first edition (1904) of his *Leçons* and this method is followed by Burkill (1953) in his Cambridge Tract. In this method geometry reigns supreme and the integral $\int_a^b f(x)\,dx$ of a non-negative function $f(x)$ is defined as the two-dimensional measure of the set of points (x, y) such that $(a \leqslant x \leqslant b, \; 0 \leqslant y \leqslant f(x))$.

In the second method the integral is defined directly and the measure of a set of points E is then defined as the integral of their 'indicator' $\alpha(x, E)$, i.e. the function that is equal to unity if $x \in E$ or to zero if $x \notin E$.

There are various techniques for a direct definition of the Lebesgue integral and we briefly refer to three of these:

(i) the method of monotone sequences invented by W. H. Young (1910) and expounded by L. C. Young (1927);

(ii) the modification of the Darboux–Riemann method described by Saks (1937, p. 3) and employed by Williamson (1962, p. 39);

853146 X   B

(iii) the use of sequences of step functions $\{\phi_n(x)\}$, with some generalized type of convergence. This is the method described by Riesz and Nagy (1953) and employed by Ingleton (1965).

Each of these methods has its own advantages, and we may apply to them the words of Kipling:

> There are nine and sixty ways of constructing tribal lays,
> And-every-single-one-of-them-is-right.

2   The concept of an integral

2.1. Introduction

In the classical treatises on the various branches of mathematics
the fundamental definitions and axioms are enunciated at the
very beginning and followed by a systematic explanation of their
logical consequences. But an introductory account of the
Lebesgue theory cannot exhibit this classical perfection, for the
fundamental definition of the Lebesgue integral is not a datum
to be unquestionably accepted but a quaesitum that has to be
achieved. The fact that we know the name of the entity—'the
Lebesgue integral'—that we have to discover must not blind us
to the fact that, at the beginning of our search, we do not know
anything more about it, except that we hope it will prove to be
a generalization of the integral that we have met in elementary
calculus. In this puzzling and paradoxical situation how can we
plan a systematic investigation?

The answer is provided by the distinction between 'constructive' and 'descriptive' definitions. Our task is to develop a
constructive definition of the Lebesgue integral that will guaran-
tee its real existence in the world of mathematics. The constructive definition will be the end of our search. But at the very
beginning we can give a descriptive definition of the Lebesgue
integral by enumerating some of the properties that it must
possess. These properties will be the most general and fundamental properties of the integrals that we have encountered
in elementary analysis. We shall find that these properties,
together with the two techniques of bracketing and of monotony,
almost inevitably decide the path that leads to a constructive
definition of the Lebesgue integral.

We therefore begin by disengaging the general concept of an
integral from the material provided by the elements of the
differential and integral calculus. In fact, elementary calculus

does provide two definitions of an integral, viz. as a 'primitive' and as an area.

By examining the concept of a primitive we shall obtain a descriptive definition of an integral and by examining the concept of area we shall see, in general terms, how a constructive definition can be achieved.

2.2. 'Primitives'

A primitive is a correlative of a derivative, i.e. if two real functions, $\phi(x)$ and $f(x)$, of a real variable x, defined and bounded in the interval $(a,b) = \{x: a < x < b\}$, are so related that

$$\frac{\phi(x+h)-\phi(x)}{h} \to f(x) \quad \text{as } h \to 0$$

for all values of x and $x+h$ in the given interval, then $f(x)$ is the derivative of $\phi(x)$ and $\phi(x)$ is a primitive of $f(x)$. Any other primitive of $f(x)$ differs from $\phi(x)$ only by an additive constant c, and has the form $\phi(x)+c$.

If the primitive function $\phi(x)$ is prescribed, and if it has been chosen from the severely restricted class of functions that do possess derivatives, then the definition given above does prescribe a definite limiting process for calculating $f(x)$ (in principle) to any prescribed degree of accuracy. The definition is therefore 'constructive'. But if it is the derivative $f(x)$ that is prescribed, then the definition is purely 'descriptive' and provides no determinate means of calculating the primitive $\phi(x)$, which in fact is usually found (when it exists!) by ingenious artifices and patient experimentation, aided by a well-stocked memory of lists of elementary functions and their derivatives. However, the familiar functions

$$f(x) = \exp(-x^2) \quad \text{or} \quad f(x) = x^{-1}\sin x$$

provide examples of integrands whose primitives cannot be expressed by any finite combination of elementary functions.

The relation of a primitive to a derivative is essentially a 'local' property, i.e. the numerical value of the derivative $f(x)$ at a specified point $x = \xi$ depends only on the values of the primitive $\phi(x)$ in an arbitrarily small neighbourhood $\xi-\epsilon < x < \xi+\epsilon$ of the point ξ. In more technical language, differentiability at a point $x = \xi$ is a local property of a function $f(x)$ because it is a property of the 'restriction' of $f(x)$ to any neighbourhood of $x = \xi$, i.e. of any function $f(x, \epsilon)$ such that $f(x, \epsilon) = f(x)$ when $\xi - \epsilon < x < \xi + \epsilon$, for some $\epsilon > 0$.

The fundamental global properties of the differential relation are the following, which we designate by (N), (L), and (P).

(N) The integral of the unit function, $f(x) \equiv 1$, over the interval $[a, b]$ is $b - a$, i.e. the function $\phi(x) = x - a$ is a primitive of the unit function $f(x) = 1$. This property is often described as the *Lebesgue normalizing condition*.

(L) If $\phi_1(x)$ and $\phi_2(x)$ are respectively primitives of $f_1(x)$ and $f_2(x)$ in the same interval $[a, b]$ then $c_1 \phi_1(x) + c_2 \phi_2(x)$ is a primitive of $c_1 f_1(x) + c_2 f_2(x)$ in the same interval, for all real numbers c_1 and c_2.

(P) If $\phi(x)$ is a primitive of $f(x)$ in an interval $[a, b]$, and if $f(x)$ is non-negative in this interval, then

$$\phi(x) \geqslant \phi(a).$$

The outstanding question that remains for investigation is to investigate the *continuity* of the differential relation. We can in fact prove that if the sequence of derivatives $\{\phi'_n(x)\}$, $(n = 1, 2, ...)$ converges *uniformly* in a closed interval $[a, b]$ to a derivative $\phi'(x)$ as n tends to infinity, then the corresponding sequence of primitives $\{\phi_n(x) - \phi_n(a)\}$ converges to $\phi(x) - \phi(a)$. This is a very restricted species of continuity for it requires that

(i) the limit of the sequence $\{\phi'_n(x)\}$ should be itself a derivative $\phi'(x)$, and that

(ii) the convergence of $\phi'_n(x)$ to $\phi'(x)$ should be uniform.

At the present stage of our investigation we cannot foresee how much these conditions can be relaxed in a descriptive definition of an integral.

We shall in fact establish three different (but related) conditions, which are each sufficient to ensure that

$$\lim_{n \to \infty} \int_a^b f_n(x) \, dx = \int_a^b f(x) \, dx,$$

where $\{f_n(x)\}$ is a sequence of integrable functions with limit function $f(x)$, viz.

(i) the condition of 'bounded convergence', i.e.

$$|f_n(x)| < K$$

for each x in $[a, b]$ and for each n, K being a constant independent of x and n (Theorem 8.7.7);

(ii) the condition of 'monotone convergence', i.e.

$$0 \leqslant f_n(x) \leqslant f_{n+1}(x)$$

for each x in $[-\infty, \infty]$ and for each n (Theorem 9.4.2);

(iii) the condition of 'dominated convergence', i.e.

$$|f_n(x)| < \psi(x)$$

for each x in $[-\infty, \infty]$ and for each n, $\psi(x)$ being integrable over $[-\infty, \infty]$ (Theorem 9.3.7).

(The theorems quoted are even more general, for they refer to integrals over measurable sets of points rather than over intervals.)

We are now in a position to give a descriptive definition of an integral. The definite integral of $f(x)$ over an interval $[a, b]$ must be (when it exists!) a real number that depends upon the values assumed by $f(x)$ in this interval. It is therefore a 'functional'. Our investigation of the properties of a primitive suggests that the definite integral of $f(x)$ over an interval $[a, b]$ should be a positive, linear, functional satisfying the Lebesgue normalizing condition. Since the properties (P), (L), and (N) are possessed by the primitives of derivatives, these properties are consistent with one another and can be taken as a descriptive definition of an integral.

2.3. Areas

In elementary analysis a constructive definition of the concept of 'area' is obtained by the method of 'exhaustion' invented by the Greek mathematicians Eudoxus and Archimedes.

The method is most simply described by considering a bounded, positive, non-decreasing function $f(x)$ defined in an

interval $[a, b]$, and the region R in the (x, y)-plane specified by the relations
$$a \leqslant x \leqslant b, \qquad 0 \leqslant y \leqslant f(x).$$

The function $f(x)$ is *not* assumed to be continuous.

(Naturally the same method is applicable to bounded, positive, non-increasing functions.)

We divide the interval $[a, b]$ by a finite number of points
$$a = x_0 < x_1 < x_2 < \ldots < x_n = b,$$

and define two step functions,
$$\left. \begin{aligned} \lambda(x) &= f(x_p) \\ \mu(x) &= f(x_{p+1}) \end{aligned} \right\} \quad \begin{aligned} &\text{if } x_p \leqslant x < x_{p+1} \\ &\text{and } p = 0, 1, 2, \ldots, n-1. \end{aligned}$$

Then, since $f(x)$ is non-decreasing,
$$\lambda(x) \leqslant f(x) \leqslant \mu(x).$$

A glance at Fig. 1 will show that the regions of the (x, y)-plane specified by the relations
$$a \leqslant x \leqslant b, \quad 0 \leqslant y \leqslant \lambda(x)$$

and
$$a \leqslant x \leqslant b, \quad 0 \leqslant y \leqslant \mu(x)$$

are bounded by two polygons, one inscribed and the other escribed to the region 'under the curve', i.e. the region
$$a \leqslant x \leqslant b, \quad 0 \leqslant y \leqslant f(x).$$

By definition the integrals
$$\int_a^b \lambda(x) \, \mathrm{d}x \quad \text{and} \quad \int_a^b \mu(x) \, \mathrm{d}x$$

are the areas of the inscribed and escribed polygons respectively, i.e.
$$\int_a^b \lambda(x) \, \mathrm{d}x = \sum_{p=0}^{n-1} (x_{p+1} - x_p) f(x_p),$$

$$\int_a^b \mu(x) \, \mathrm{d}x = \sum_{p=0}^{n-1} (x_{p+1} - x_p) f(x_{p+1}).$$

Hence
$$\int_a^b \mu(x) \, \mathrm{d}x - \int_a^b \lambda(x) \, \mathrm{d}x = \sum_{p=0}^{n-1} (x_{p+1} - x_p)\{f(x_{p+1}) - f(x_p)\},$$

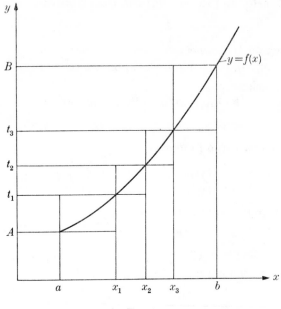

and, if the maximum length of the intervals (x_p, x_{p+1}) for $p = 0, 1, 2, ..., n-1$, is ϵ, then

$$0 \leqslant x_{p+1} - x_p \leqslant \epsilon$$

and

$$\int_a^b \mu(x)\,dx - \int_a^b \lambda(x)\,dx \leqslant \epsilon \sum_{p=0}^{n-1} \{f(x_{p+1}) - f(x_p)\}$$
$$= \epsilon\{f(b) - f(a)\}.$$

Thus, in the terminology of § 1.4, the function $f(x)$ is bracketed by the step functions $\lambda(x)$ and $\mu(x)$ with a tolerance $\epsilon\{f(b) - f(a)\}$, which can be made arbitrarily small by sufficiently increasing the number of sub-intervals into which the interval $[a, b]$ is divided.

Now

$$0 \leqslant \int_a^b \lambda(x)\,dx \leqslant \int_a^b \mu(x)\,dx \leqslant \sum_{p=0}^{n-1} (x_{p+1} - x_p)f(b) = (b-a)f(b).$$

Hence, if we consider all possible divisions of the interval $[a, b]$, the collection of integrals $\Lambda = \int_a^b \lambda(x)\, \mathrm{d}x$ has an upper bound, while the collection of integrals $M = \int_a^b \mu(x)\, \mathrm{d}x$ has a lower bound. Therefore the integrals Λ have a supremum or least upper bound, sup Λ, while the integrals M have an infimum or greatest lower bound, inf M. Also

$$0 \leqslant \inf M - \sup \Lambda \leqslant \epsilon\{f(b) - f(a)\}$$

for all $\epsilon > 0$. Therefore

$$\sup \Lambda = \inf M = A, \text{ say.}$$

Hence there is a unique number A such that, for any division of the interval $[a, b]$, $\qquad \Lambda \leqslant A \leqslant M,$

and this number is therefore defined to be the integral $\int_a^b f(x)\, \mathrm{d}x$. This we may call the 'Archimedean' integral.

$$\text{Clearly} \quad (b-a)f(a) \leqslant \int_a^b f(x)\, \mathrm{d}x \leqslant (b-a)f(b),$$

so that the Archimedean integral of a monotone function satisfies the mean value theorem, and hence is a positive functional.

2.4. The Lebesgue integral

The success of the method of exhaustion applied to the Archimedean integral clearly depends upon the closeness with which the bracketing functions $\lambda(x)$ and $\mu(x)$ approximate to the integrand $f(x)$. In fact since

$$\lambda(x) = f(x_p) \leqslant f(x) \leqslant f(x_{p+1}) = \mu(x)$$

in the interval $x_p \leqslant x < x_{p+1}$, it follows that

$$\mu(x) - \lambda(x) = f(x_{p+1}) - f(x_p),$$

i.e. the variation of $f(x)$ in this interval.

If $f(x)$ is non-decreasing and continuous in $[a, b]$ then, for any prescribed tolerance $\epsilon > 0$ we can choose a division of the interval such that

$$f(x_{p+1}) - f(x_p) < \epsilon \quad \text{for } p = 0, 1, 2, ..., n-1,$$

whence $\qquad\qquad \mu(x)-f(x) < \epsilon$

and $\qquad\qquad\qquad f(x)-\lambda(x) < \epsilon$

at all points x in $[a, b]$.

Again, if $f(x)$ is continuous, but not necessarily monotone, in $[a, b]$ we can define as bracketing functions the representations

$$\left.\begin{array}{l} \lambda(x) = \inf f(x) \\ \mu(x) = \sup f(x) \end{array}\right\} \text{ in each interval } x_p \leqslant x < x_{p+1}.$$

Then, since $f(x)$ is uniformly continuous, for any arbitrary tolerance $\epsilon < 0$ we can choose a division of the interval $[a, b]$ such that

$$\sup f(x) - \inf f(x) < \epsilon$$

in each sub-interval $x_p \leqslant x < x_{p+1}$, $p = 0, 1, 2,..., n-1$.

If, however, $f(x)$ is neither monotone nor continuous, we are faced with the embarrassing possibility that it may be wildly oscillatory in value and may have an infinity of maxima and minima in each sub-interval of $[a, b]$. The existence of such functions is assured by the example

$$f(x) = \sum_{n=1}^{\infty} \frac{1}{n^2} \sin \frac{1}{x-z_n}$$

if x is irrational, or $f(x) = 0$ if x is rational, where $z_1, z_2,...$ is some enumeration of the rational numbers between 0 and 1.

In these circumstances bracketing functions based on a division of the domain $[a, b]$ of $f(x)$ would be useless, and we proceed to consider a new species of bracketing function based on a division of the *range* $[A, B]$ of $f(x)$—this is one of the major contributions of Lebesgue to the theory of integration.

We divide the range $[A, B]$ of the function $f(x)$ by a finite number of points

$$A = t_0 < t_1 < t_2 < ... < t_n = B$$

and introduce the function

$$\psi_p(x) = \begin{cases} 1 & (t_p < f(x) \leqslant t_{p+1}) \\ 0 & (f(x) \leqslant t_p \text{ or } t_{p+1} < f(x)) \end{cases}$$

so that $\qquad\qquad \sum_{p=0}^{n-1} \psi_p(x) = 1 \quad \text{if } a \leqslant x \leqslant b.$

We define two bracketing functions,

$$\lambda(x) = \sum_{p=0}^{n-1} t_p \, \psi_p(x),$$

$$\mu(x) = \sum_{p=0}^{n-1} t_{p+1} \psi_p(x).$$

Then, if ϵ is the maximum length of the intervals (t_p, t_{p+1}) for $p = 0, 1, 2, ..., n-1$,

$$0 \leqslant t_{p+1} - t_p \leqslant \epsilon,$$

and

$$\mu(x) - \lambda(x) \leqslant \sum_{p=0}^{n-1} (t_{p+1} - t_p) \psi_p(x)$$

$$< \epsilon \sum_{p=0}^{n-1} \psi_p(x) = \epsilon.$$

Now

$$\lambda(x) \leqslant f(x) \leqslant \mu(x).$$

Hence the arbitrary function $f(x)$ is uniformly approximated by the bracketing functions $\lambda(x)$ and $\mu(x)$ over the whole interval $[a, b]$.

Moreover, if we possessed a definition of integration applicable to the bracketing functions we could conclude at once that

$$\int_a^b \mu(x) \, dx - \int_a^b \lambda(x) \, dx = \int_a^b \{\mu(x) - \lambda(x)\} \, dx \leqslant \epsilon(b-a).$$

We could then define the integral of $f(x)$ over the interval $[a, b]$ as

$$\int_a^b f(x) \, dx = \inf \int_a^b \mu(x) \, dx = \sup \int_a^b \lambda(x) \, dx$$

for all bracketing functions of the type defined above. This is the Lebesgue integral!

It thus appears that the problem of integrating a bounded function $f(x)$ over a bounded interval $[a, b]$ would be solved if we could integrate the bracketing functions $\lambda(x)$ and $\mu(x)$, and that these bracketing functions would be integrable if we could integrate the functions $\psi_p(x)$.

This reconaissance of the problem of integration, due to Lebesgue (1928, chap. vii) shows that the essence of the matter is the integration of functions, such as $\psi_p(x)$, which can take only two values, viz 0 and 1. We therefore proceed to make a preliminary examination of the geometrical significance of the integrals $\int\limits_a^b \psi_p(x)\,\mathrm{d}x$.

2.5. Lebesgue measure

(i) In the simplest case, when $f(x)$ is a non-decreasing function of x in $[a,b]$, then to any number t_p in the range $[A, B]$ of $f(x)$ there corresponds a unique number x_p such that

$$a \leqslant x_p \leqslant b$$

and
$$\begin{cases} f(x) \leqslant t_p & (x \leqslant x_p), \\ f(x) > t_p & (x > x_p). \end{cases}$$

Then
$$t_p < f(x) \leqslant t_{p+1} \quad \text{if } x_p < x \leqslant x_{p+1}.$$

Hence
$$\int\limits_a^b \psi_p(x)\,\mathrm{d}x = \int\limits_{x_p}^{x_{p+1}} 1\,.\,\mathrm{d}x = x_{p+1} - x_p.$$

Therefore the Lebesgue bracketing functions

$$\sum_{p=0}^{n-1} t_p\,\psi_p(x) \quad \text{and} \quad \sum_{p=0}^{n-1} t_{p+1}\,\psi_p(x)$$

have exactly the same integrals

$$\sum_{p=0}^{n-1} f(x_p)(x_{p+1}-x_p) \quad \text{and} \quad \sum_{p=0}^{n-1} f(x_{p+1})(x_{p+1}-x_p)$$

as the step functions that we introduced as bracketing functions in § 2.3.

THEOREM 2.5.1. *If $f(x)$ is a non-decreasing function, it possesses a Lebesgue integral, which is exactly the same as the 'Archimedean' integral of* § 2.3.

(ii) To discuss the integration of $\psi_p(x)$ when $f(x)$ is continuous but not monotone in $[a,b]$ it is more convenient to introduce the function
$$\alpha(x,t) = \begin{cases} 1 & (t < f(x)), \\ 0 & (f(x) \leqslant t). \end{cases}$$

Then
$$\psi_p(x) = \alpha(x,t_p) - \alpha(x,t_{p+1}).$$

The advantage of this transformation is that it is easily proved that the set of points at which $\alpha(x,t_p) = 1$ is 'open', i.e. it consists of an enumerable collection of open intervals (a_k, b_k), $(k = 1, 2, ...)$. Thus $\int_{a_k}^{b_k} \alpha(x, t_p) \, dx = b_k - a_k$, and $\int_a^b \alpha(x, t_p) \, dx$ may be suitably defined as $\sum_{k=1}^{\infty} (b_k - a_k)$, since the series of positive terms has an upper bound $(b-a)$ and is therefore convergent. (This, of course, requires proof, which we shall supply below.)

We may cite, as an example, the function

$$f(x) = \begin{cases} x \sin \dfrac{\pi}{x} & (x \neq 0), \\ 0 & (x = 0), \end{cases}$$

in the domain $(0, 1)$ and the auxiliary function

$$\alpha(x, 0) = \begin{cases} 1 & (0 < f(x)), \\ 0 & (f(x) \leqslant 0). \end{cases}$$

Then $\int_0^1 \alpha(x, 0) \, dx = \frac{1}{2} - \frac{1}{3} + \frac{1}{4} - \frac{1}{5} + ... = 1 - \ln 2.$

(iii) In the general case, when $f(x)$ is neither monotone nor continuous, further research is necessary to determine under what conditions the function $\alpha(x, t)$ can be integrated. The answer is provided by the Lebesgue theory of measure that we shall develop in Chapter 7.

Assuming that the given function $f(x)$ is such that the auxiliary function $\alpha(x, t)$ is integrable with respect to x for each value of t, let

$$m(t) = \int_a^b \alpha(x, t) \, dx.$$

This function $m(t)$ will be called the '*measure function*' of $f(x)$ and will be proved to be a non-increasing function of t.

The Lebesgue bracketing functions will then have the integrals

$$\Lambda = \sum_{p=0}^{n-1} t_p \{ m(t_p) - m(t_{p+1}) \} \quad \text{and} \quad M = \sum_{p=0}^{n-1} t_{p+1} \{ m(t_p) - m(t_{p+1}) \},$$

and as before

$$\begin{aligned} 0 \leqslant M - \Lambda &\leqslant \epsilon \sum_{p=0}^{n-1} \{ m(t_p) - m(t_{p+1}) \} \\ &= \epsilon \{ m(t_0) - m(t_n) \} \\ &= \epsilon (b - a). \end{aligned}$$

The Lebesgue integral of $f(x)$ over $[a, b]$ can then be defined as

$$\int_a^b f(x)\,\mathrm{d}x = \sup \Lambda = \inf \mathbf{M}$$

for all Lebesgue bracketing functions, bracketing $f(x)$.

2.6. The structure of the Lebesgue theory of integration

The preceding survey of the problem of integration, as analysed by Lebesgue, shows that a necessary preliminary is the theory of measure.

To enable the essential features of the theory to be grasped more readily we shall give prior consideration to the theories of measure and integration in one dimension, but the terminology, theorems, and definitions will be so phrased that they can readily be interpreted in the d-dimensional theory.

We shall show that the Lebesgue integral is a positive, linear continuous functional and we shall investigate in what sense the indefinite Lebesgue integral

$$\phi(x) = \int_a^x f(t)\,\mathrm{d}t \quad (a \leqslant x \leqslant b)$$

can be regarded as a primitive of $f(x)$.

However, there is a chapter in the theory of Lebesgue integration that is even simpler than the one-dimensional theory. This is the chapter that discusses sets of points of zero measure. This theory of 'null sets' is not only of extreme simplicity, but it permeates the whole of Lebesgue theory. In particular, it is indispensable for the discussion of the differentiability of the indefinite integral

$$\phi(x) = \int_a^x f(t)\,\mathrm{d}t.$$

We shall therefore begin our discussion of the measure theory with a chapter on sets of zero measure (Chapter 5).

Before we proceed further to discuss the Lebesgue theory in one dimension it is convenient to give a rather more formal exposition of the two main instruments employed in the theory—the method of monotony and the method of bracketing. Also, in order to express the theory of measure in analytical form (rather than in the usual geometrical form) we shall intercalate a chapter on the theory and use of the indicator function $\alpha(x, t)$.

2.7. Exercises

1. If the sequence of derivatives $\{\phi_n'(x)\}$ is non-negative, monotonic in x, monotonic in n, and converges to zero as $n \to \infty$ in $[a, b]$, prove that
$$\phi_n(x) - \phi_n(a) \to 0 \quad \text{as } n \to \infty \quad \text{(Denjoy)}.$$

2. If the function $f(x)$ possesses a derivative $f'(x)$ at each point of the interval $a \leqslant x \leqslant b$, $f'(a) = \alpha$, $f'(b) = \beta$, $\beta \neq \alpha$, and γ lies between α and β, then there is a point c between a and b such that $f'(c) = \gamma$ (Darboux).

3. Deduce from the properties (N), (L), (P) of the differential relation (§ 2.2) that, if $\phi(x)$ is a primitive of $f(x)$ in the interval $[a, b]$, and if
$$L \leqslant f(x) \leqslant M \quad \text{for } a \leqslant x \leqslant b,$$
then $L(b-a) \leqslant \phi(b) - \phi(a) \leqslant M(b-a)$.
  Hence deduce Rolle's theorem.

4. If $f(x)$ is a continuous non-decreasing function in $[a, b]$ and c is any constant, prove that
$$\int_a^b \{c - f(x)\}\, \mathrm{d}x = c(b-a) - \int_a^b f(x)\, \mathrm{d}x.$$

3 The techniques of Lebesgue theory

3.1. The starting-point

In seeking for the widest possible generalization of the concept of an integral there are two directions that we may follow—we may try to generalize the concept of a primitive or we may try to generalize the concept of area. The first method has been followed by Perron, Ward, and Henstock, the second method by Riemann, Borel, Young, and Lebesgue.

In following the method of Lebesgue the starting-point is necessarily the concept of the area of a rectangle and its immediate extension to the integrals of step functions such as the bracketing functions of § 2.3. To change the metaphor, no other material is available for the construction of the Lebesgue integral than the integrals of step functions.

It follows that the only techniques available for the process of construction are those already employed in constructing the functions of analysis from step functions. These are the familiar algebraical techniques of addition, multiplication, and their inverses, together with the analytical techniques of limiting processes.

3.2. The method of bracketing

The two techniques of generalization characteristic of Lebesgue theory are the method of bracketing and the method of monotonic convergence.

The method of bracketing has been briefly discussed in § 1.4 as a technique for extending the concept of integration, but it is interesting to indicate the very extensive class of functionals to which it can be applied and to expose the inherent restrictions in this method. Briefly we shall show that the method of bracketing can be applied to the class of functionals $I(f)$, which are characterized by the property that, if $f(x) \geqslant g(x)$, then

$I(f) \geqslant I(g)$, but that bracketing is necessarily a 'closure' operation that can only be applied once to extend a given class of functionals.

DEFINITION 3.2.1. A functional $I(f)$, defined for a set F of functions $f(x)$ defined in an interval ($a \leqslant x \leqslant b$), is said to be 'monotonic' if $I(f) \geqslant I(g)$ whenever $f(x) \geqslant g(x)$ for all x in $[a, b]$, and $f(x)$, $g(x)$ belong to the class F.

DEFINITION 3.2.2. A function $\phi(x)$ is said to be 'bracketed' by two functions $\lambda(x, \epsilon)$, $\mu(x, \epsilon)$ (with tolerance $\epsilon > 0$) if λ and μ belong to the domain F of a monotonic functional $I(f)$ and if

$$\lambda(x, \epsilon) \leqslant \phi(x) \leqslant \mu(x, \epsilon),$$
$$I(\mu) - I(\lambda) \leqslant \epsilon.$$

THEOREM 3.2.1. *If, for each prescribed tolerance $\epsilon > 0$, the function $\phi(x)$ is bracketed by two functions $\lambda(x, \epsilon)$, $\mu(x, \epsilon)$ belonging to the domain F of a functional $I(\lambda)$ then the supremum of $I(\lambda)$ and the infimum of $I(\mu)$ both exist and are equal.*

Let λ_0, μ_0 be any fixed pair of bracketing functions. Then

$$\lambda_0 \leqslant \phi \leqslant \mu_0, \quad \lambda \leqslant \phi \leqslant \mu.$$

Hence $\qquad\qquad \lambda \leqslant \mu_0, \quad \lambda_0 \leqslant \mu,$

and $\qquad\qquad I(\lambda) \leqslant I(\mu_0), \quad I(\lambda_0) \leqslant I(\mu).$

Thus the numbers $I(\lambda)$ are bounded above, and the numbers $I(\mu)$ are bounded below. Hence the numbers $I(\lambda)$ have a supremum or least upper bound, $\sup I(\lambda)$, and the numbers $I(\mu)$ have an infimum or greatest lower bound, $\inf I(\mu)$.

For any prescribed tolerance $\epsilon \geqslant 0$ there exist bracketing functions λ and μ such that

$$I(\mu) - I(\lambda) \leqslant \epsilon.$$

Hence $\qquad \inf I(\mu) - \sup I(\lambda) \leqslant \epsilon, \quad \text{for all } \epsilon > 0,$

and therefore $\qquad \sup I(\lambda) = \inf I(\mu).$

DEFINITION 3.2.3. With the notation and terminology of Theorem 3.2.1, the 'bracketed' functional $I^*(\phi)$ is defined to be

$$I^*(\phi) = \sup I(\lambda) = \inf I(\mu).$$

THEOREM 3.2.2. *If $\phi(x)$ belongs to the domain F of the functional $I(\phi)$, then*
$$I^*(\phi) = I(\phi).$$

For
$$\sup I(\lambda) = I(\phi) = \inf I(\mu).$$

Hence the bracketed functional $I^*(\phi)$ of a function of the class F is equal to the functional $I(\phi)$. Thus the bracketed functional is a valid and self-consistent extension of the original function, which we may call the *bracketed extension*.

It is important to notice at once the intrinsic limitation of the method of bracketing. Let F be the domain of a monotonic functional $I(f)$ and Φ the domain of the bracketed functional $I^*(\phi)$. Is it possible to construct a further extension of the original functional by the use of functions $\psi(x)$ which are bracketed by functions from the class Φ? The answer is in the negative.

THEOREM 3.2.3. *If, for any prescribed tolerance $\epsilon > 0$, a function $\psi(x)$ is bracketed by two functions $\phi_1(x)$ and $\phi_2(x)$ of the set Φ, then $\psi(x)$ also belongs to the set Φ.*

For there exist bracketing functions ϕ_1, ϕ_2 of set Φ such that
$$\phi_1 \leqslant \psi \leqslant \phi_2,$$
and
$$I^*(\phi_2) - I^*(\phi_1) \leqslant \epsilon.$$

Also there exist bracketing functions λ_1, μ_1, λ_2, μ_2 of set F such that
$$\lambda_1 \leqslant \phi_1 \leqslant \mu_1, \qquad \lambda_2 \leqslant \phi_2 \leqslant \mu_2$$
and
$$I^*(\mu_1) - I^*(\lambda_1) \leqslant \epsilon, \qquad I^*(\mu_2) - I^*(\lambda_2) \leqslant \epsilon.$$

Note also that
$$I^*(\lambda_2) \leqslant I^*(\phi_2) \quad \text{and} \quad I^*(\phi_1) \leqslant I^*(\mu_1).$$

Then
$$\lambda_1 \leqslant \psi \leqslant \mu_2$$
and

$$I^*(\mu_2) - I^*(\lambda_1)$$
$$= \{I^*(\mu_2) - I^*(\phi_2)\} + \{I^*(\phi_1) - I^*(\lambda_1)\} + \{I^*(\phi_2) - I^*(\phi_1)\}$$
$$\leqslant \{I^*(\mu_2) - I^*(\lambda_2)\} + \{I^*(\mu_1) - I^*(\lambda_1)\} + \{I^*(\phi_2) - I^*(\phi_1)\}$$
$$\leqslant 3\epsilon.$$

Thus ψ is bracketed with arbitrary tolerance 3ϵ, by functions λ_1 and μ_2 of the class F. Hence ψ belongs to the same set Φ as the functions ϕ_1, ϕ_2, and we have not succeeded in making any further extension of the domain of the bracketed functional $I^*(\phi)$.

The operation of bracketing is thus similar to the closure operation in point-set topology in as much as both are idempotent operations, i.e. the repetition of the operation produces no further extension of the set to which they are applied.

3.3. The Riemann integral

As we have shown in § 2.3 bracketing by step functions furnishes the integral of bounded, monotonic functions. In fact the scope of this technique is much larger and it provides the simplest definition of the Riemann integral.

As before we divide the integral $[a, b]$ by a finite number of points $$a = x_0 < x_1 < x_2 < ... < x_n = b.$$
Let L_p and M_p be the greatest lower bound and the least upper bound respectively of a function $f(x)$ in the interval $x_p \leqslant x < x_{p+1}$, for $p = 0, 1, 2, ..., n-1$. Let F_p be any number in the range $[L_p, M_p]$. Then the sum

$$S = \sum_{p=0}^{n-1} F_p(x_{p+1} - x_p)$$

is a Riemann approximation to the integral of $f(x)$ in $[a, b]$.

Let $\epsilon = \max(x_{p+1} - x_p)$ for $p = 0, 1, 2, ..., n-1$. As $\epsilon \to 0$, $n \to \infty$ and the corresponding sums S may converge to a limit. If so, this limit is the Riemann integral $R \int_a^b f(x)\, \mathrm{d}x$.

It is, however, clear that $f(x)$ is bracketed by the step function

$$\left.\begin{array}{l} \lambda(x) = L_p \\ \mu(x) = M_p \end{array}\right\} \text{ if } x_p \leqslant x < x_{p-1},$$

and that $\quad \lambda(x) \leqslant f(x) \leqslant \mu(x) \quad$ if $a \leqslant x < b$,

while $\quad \Lambda = \int_a^b \lambda(x)\, \mathrm{d}x = \sum_{p=0}^{n-1} L_p(x_{p+1} - x_p),$

$$\mathrm{M} = \int_a^b \mu(x)\, \mathrm{d}x = \sum_{p=0}^{n-1} M_p(x_{p+1} - x_p),$$

whence

$$0 \leqslant M - \Lambda \leqslant \sum_{p=0}^{n-1} (M_p - L_p)(x_{p+1} - x_p) \leqslant \epsilon \sum_{p=0}^{n-1} (M_p - L_p).$$

As before, the collection of numbers M has a greatest lower bound, inf M, and the collection of numbers Λ has a least upper bound, sup Λ. These numbers are the upper and lower Darboux integrals of $f(x)$ over $[a, b]$.

Now $$L_p \leqslant F_p \leqslant M_p$$
and $$\Lambda \leqslant S \leqslant M.$$

Hence if the upper and lower Darboux integrals are equal, then the Riemann approximations S must converge to the common value of sup Λ and inf M. Conversely we can always choose the numbers F_p so that $L_p = F_p$ and $\Lambda = S$ or so that $M_p = F_p$ and $M = S$. Thus the equality of the upper and lower Darboux integrals is a necessary and sufficient condition for the existence of the Riemann integral.

We note without proof that any one of the following conditions is sufficient to ensure the existence of the Riemann integral of a bracketed function $f(x)$:

(1) $f(x)$ is continuous (and therefore uniformly continuous) in $[a, b]$,
(2) $f(x)$ has only an enumerable set of discontinuities in $[a, b]$,
(3) $f(x)$ has bounded variation in $[a, b]$,

and, of course, it is sufficient if $f(x)$ is monotone in $[a, b]$.

The simplicity and naturalness of the Riemann integral leave little to be desired, and it is easy to show that it is a positive, absolute, and linear functional (see Exercise 3). But unfortunately it is not continuous, i.e. if each of the functions $f_n(x)$ in a convergent sequence is integrable by Riemann's method, it is not necessarily true that the limit function $f(x)$ is also integrable by the same method.

Consider, for example, any enumeration $\{z_n\}$ ($n = 1, 2, ...$) of the rational numbers between 0 and 1, and let

$$f_n(x) = \begin{cases} 1 & (x = z_1, z_2, ..., z_n), \\ 0 & \text{otherwise.} \end{cases}$$

Then in any subdivision of the interval $[0, 1]$,

$$L_p = 0, \qquad M_p = 1.$$

and $$\Lambda = 0, \qquad M = 1.$$

The limit function $f(x)$ has the value unity if x is rational and the value zero if x is irrational. The upper and lower Darboux integrals are respectively 1 and 0, whence $f(x)$ has no Riemann integral.

To be just to the Riemann integral, which has played so great a part in nineteenth-century analysis, it must be stated that it does possess a certain restricted continuity, in the sense that if the functions $\{f_n(x)\}$ are each integrable by Riemann's method, if the sequence $\{f_n(x)\}$ converges uniformly to the limit function $f(x)$, and if $f(x)$ is also integrable by Riemann's method, then

$$\int_a^b f_n(x) \, \mathrm{d}x \to \int_a^b f(x) \, \mathrm{d}x \quad \text{as } n \to \infty.$$

But this is far too restrictive a condition even in the theory of Fourier series, and having saluted the memory of Riemann's integral, we therefore pass on to our main topic—the Lebesgue integral.

3.4. Monotone sequences

Monotone sequences play an important role in Lebesgue theory, and we therefore summarize their main properties in this section. It is convenient to reword the traditional definitions of limiting points in order to get a more compact description.

DEFINITION 3.4.1. 'Almost all' numbers of an enumerable sequence $\{u_n\}$ $(n = 1, 2,...)$ are said to possess a property P if there is only a *finite* number that do *not* possess the property.

If $\{u_n\}$ is any sequence bounded below, then there exists an infimum or greatest lower bound $\inf u_n = \lambda$ such that

(i) $u_n \geqslant \lambda$ for all n, and
(ii) there is at least one number of the sequence in any *closed* interval $[\lambda, \lambda+\epsilon]$ where $\epsilon > 0$.

There is also a lower limit $\Lambda = \liminf u_n$ such that, for each tolerance $\epsilon > 0$,

  (i) $u_n \geqslant \Lambda - \epsilon$ for almost all n, and
  (ii) there is at least one number of the sequence less than $\Lambda + \epsilon$.

But if $\{u_n\}$ is a monotone, non-increasing sequence, bounded below,

i.e.   if $u_n \geqslant u_{n+1}$, for all n,

then there exists a unique limit, $l = \lim u_n$, such that

  (i) $u_n \geqslant l$ for all n,   and
  (ii) $u_n < l + \epsilon$ for almost all n.

Similarly if $\{u_n\}$ is any sequence bounded above, then there exists a supremum or least upper bound, $\sup u_n = \mu$, such that

  (i) $u_n \leqslant \mu$ for all n, and
  (ii) there is at least one number of the sequence in any *closed* interval $[\mu - \epsilon, \mu]$.

There is also an upper limit $M = \limsup u_n$ such that, for each tolerance $\epsilon > 0$,

  (i) $u_n \leqslant M + \epsilon$ for almost all n, and
  (ii) there is at least one number of the sequence greater than $M - \epsilon$.

But if $\{u_n\}$ is a monotone, non-decreasing sequence, bounded above,

i.e.   if $u_n \leqslant u_{n+1}$ for all n,

then there exists a unique limit, $m = \lim u_n$, such that

  (i) $u_n \leqslant m$ for all n,   and
  (ii) $u_n > m - \epsilon$ for almost all n.

The study of the convergence of any bounded sequence $\{u_n\}$ can be reduced to the study of two monotonic sequences by the use of the associated 'peak and chasm' sequences.

DEFINITION 3.4.2. Let

$$\pi_n = \sup u_p \quad \text{for } p \geqslant n,$$
$$\chi_n = \inf u_p \quad \text{for } p \geqslant n.$$

Then $\{\pi_n\}$ and $\{\chi_n\}$ are the peak and chasm sequences associated with the original sequence $\{u_n\}$.

The peak sequence is non-increasing and the chasm sequence is non-decreasing. Their limits

$$\lim \pi_n = \lim \sup u_n = \overline{\lim}\, u_n$$

and $$\lim \chi_n = \lim \inf u_n = \underline{\lim}\, u_n$$

as $n \to \infty$, are the 'upper' and 'lower' limits of the sequence $\{u_n\}$.

The necessary and sufficient condition that the original sequence $\{u_n\}$ should converge to a unique limit λ is that

$$\lim \inf u_n = \lim \sup u_n = \lim u_n = \lambda.$$

All these considerations apply of course to a sequence $\{u_n\}$ defined as $u_n = f_n(x)$, i.e. the value of the functions $f_n(x)$ at a specific point x.

For example, $u_n(x)$ and $v_n(x)$, the real and imaginary parts of the partial sums

$$\sum_{p=0}^{n} e^{ipx} \quad (0 < x < 2\pi)$$

of the Fourier series of a Dirac delta function, are represented on the Argand diagram by points on a circle of centre $\frac{1}{2} + \frac{1}{2}i \cot \frac{1}{2}x$ and of radius $\operatorname{cosec} \frac{1}{2}x$. Hence, if $x/2\pi$ is irrational, we easily find that
$$\lim \sup u_n = \tfrac{1}{2} + \operatorname{cosec} \tfrac{1}{2}x,$$
$$\lim \inf u_n = \tfrac{1}{2} - \operatorname{cosec} \tfrac{1}{2}x,$$
$$\lim \sup v_n = \tfrac{1}{2} \cot \tfrac{1}{2}x + \tfrac{1}{2} \operatorname{cosec} \tfrac{1}{2}x,$$
$$\lim \inf v_n = \tfrac{1}{2} \cot \tfrac{1}{2}x - \tfrac{1}{2} \operatorname{cosec} \tfrac{1}{2}x.$$

(If $x/2\pi$ is rational, the corresponding results are slightly more complicated.)

3.5. Infinite integrals

In the Lebesgue theory the upper and lower bracketing functions are bounded, whence the definition of bracketed integrals (§ 3.2) is obviously restricted to *bounded* functions. In the Lebesgue theory the extension of the definition of integration to unbounded functions and to infinite intervals of integration is made particularly simple because we have to consider only integrals of *non-negative* functions, as will appear later in § 9.2. As a result the extension of the concept of integration can be made in terms of monotonic convergence.

DEFINITION 3.5.1. If $f(x)$ is a non-negative function defined for all values of x, then the 'truncated function' $f_{s,t}(x)$ is defined as

$$f_{s,t}(x) = f(x) \quad \text{if} \quad -s \leqslant x \leqslant s \quad \text{and} \quad f(x) \leqslant t,$$

or $\quad f_{s,t}(x) = t \quad \text{if} \quad -s \leqslant x \leqslant s \quad \text{and} \quad f(x) > t,$

or $\quad f_{s,t}(x) = 0 \quad \text{if} \quad |x| > s,$

s and t being positive numbers.

Thus $f_{s,t}(x)$ converges monotonically to $f(x)$ as s and t converge independently to zero.

DEFINITION 3.5.2. If each of the truncated functions $f_{s,t}(x)$ belongs to the domain of a monotonic functional $I(\;\;)$, then the 'monotonic' extension of $I(\;\;)$ for the function $f(x) = \lim f_{s,t}(x)$ is

$$I^*(f) = \lim I(f_{s,t}) \quad \text{as } s, t \to \infty.$$

If $s < \sigma$ and $t < \tau$, then

$$I(f_{s,t}) \leqslant I(f_{\sigma,\tau}),$$

whence $I(f_{s,t})$ necessarily converges to a limit (finite or infinite) as $s, t \to \infty$.

It is almost trivial to note

THEOREM 3.5.1. *In Definition* 3.5.2, *the variables* s *and* t *can be restricted to the positive integers, and moreover we may take them to be equal so that* $s = t = n$.

It is, however, important to state

THEOREM 3.5.2. *If* $f(x)$ *is bounded and defined on a bounded interval* $[a, b]$, *then* $\quad I^*(f) = I(f).$

For $f_{s,t}(x) = 0$ if $s > \max(a, b)$.

This result is necessary to establish the consistency of our definition of the monotone extension of $I(\;\;)$.

3.6. The Dini derivatives

In discussing the differentiation of an indefinite Lebesgue integral (§ 5.1) we shall have to take explicit cognisance of the fact that a continuous function does not necessarily possess a unique derivative everywhere. The elementary example

$$f(x) = |x|$$

shows that $f(x)$ may not be differentiable at the origin, where

$$\frac{f(x+h)-f(x)}{h} \to +1 \quad \text{or} \quad -1, \quad \text{for } x = 0,$$

according as h tends to zero through positive or negative values.

By various ingenious methods it is possible to construct continuous functions that are not differentiable at any rational point, or indeed at any point whatsoever, and we refer to *The theory of functions* by E. C. Titchmarsh (§§ 11.21–11.23) for an account of what has been called the 'morbid pathology' of analysis.

When a function $f(x)$ fails to possess a derivative in the ordinary sense, i.e. when the incrementary ratio

$$G(x,h) = \frac{f(x+h)-f(x)}{h}$$

does not tend to a unique limit as $|h| \to 0$, we can employ the peak and chasm functions of § 3.4 to define what are commonly called the 'Dini' derivatives, after the Italian mathematician who introduced them into analysis. There is this difference that in § 3.4 we were considering the limits of a set of numbers $f_n(x)$ which were defined for integral values of n, whereas now we are concerned with the limits of a set of numbers $G(x,h)$ which are defined for values of h in a continuous interval, $-\delta < h < \delta$.

Let $\pi(x,\delta)$ be the supremum of $G(x,h)$ and $\chi(x,\delta)$ be the infimum value of $G(x,h)$ in the domain $0 < h < \delta$. Then as $\delta \to 0$, $\pi(x,\delta)$ is non-increasing, and $\chi(x,\delta)$ is non-decreasing. Therefore $\pi(x,\delta)$ and $\chi(x,\delta)$ each tend monotonically to unique limits as $\delta \to 0$. These are upper and lower Dini derivatives of $f(x)$ on the right, usually denoted by

$$D^+f(x) = \lim_{\delta \to 0} \pi(x, \delta) = \limsup_{h \to 0} G(x, h) \quad (0 < h < \delta),$$

$$D_+f(x) = \lim_{\delta \to 0} \chi(x, \delta) = \liminf_{h \to 0} G(x, h) \quad (0 < h < \delta).$$

Similarly, the upper and lower Dini derivatives of $f(x)$ on the left are defined as

$$D^-f(x) = \limsup_{h \to 0} G(x, h) \quad (-\delta < h < 0),$$

$$D_-f(x) = \liminf_{h \to 0} G(x, h) \quad (-\delta < h < 0).$$

The Dini derivatives always exist, but their numerical value may be $+\infty$ or $-\infty$.

3.7. Exercises

A functional $I(f)$ is said to be

   (i)   positive if $f \geqslant 0$ implies that $I(f) \geqslant 0$,
   (ii) additive if $I(f_1 + f_2) = I(f_1) + I(f_2)$,
   (iii) multiplicative if $I(cf) = cI(f)$ for any constant c,
   (iv) linear if $I(c_1 f_1 + c_2 f_2) = c_1 I(f_1) + c_1 I(f_2)$, for any constants c_1, c_2,
   (v)   'absolute' if the existence of $I(f)$ implies the existence of $I(|f|)$,
   (vi) completely additive if the conditions

$$f_n \geqslant 0, \quad I(f_n) \text{ exists}, \quad \sum_{p=1}^{n} f_p \to f \text{ as } n \to \infty,$$

and $I(f)$ exists, are sufficient to ensure that

$$\sum_{p=1}^{\infty} I(f_p) = I(f).$$

1. Prove that the integral of a step function is a monotonic, positive, linear, absolute functional.

2. If $I^*(\phi)$ is the bracketed extension of $I(f)$, prove that

   (i)   if I is positive, so also is I^*,
   (ii) if I is additive, so also is I^*,
   (iii) if I is multiplicative, so also is I^*,
   (iv) if I is linear, so also is I^*,
   (v)   if I is linear and absolute, so also is I^*,
   (vi) if I is positive, linear, additive, and completely additive, so also is I^*.

3. The Riemann integral $R(f)$ can be defined as the bracketed extension of the monotone functional $I(f)$ for step functions in a bounded interval.

Examine which of the properties listed above are possessed by the Riemann integral.

4. Show that the Riemann integral over a bounded interval of non-negative, non-decreasing functions $f(x)$ is a completely additive functional (Denjoy 1941–9, p. 428), i.e. if $f_n \geqslant 0$, $f_n(x)$ is non-decreasing in x, $f_n(x)$ is uniformly bounded, then $\sum\limits_{p=1}^{n} f_p(x)$ converges to a non-decreasing function $f(x)$ as $n \to \infty$, and

$$\sum_{p=1}^{\infty} I(f_n) = I(f).$$

5. (i) If $f_n(x)$ possesses a non-negative derivative $f'_n(x)$ for each integer n and each x in (a, b), and if

$$s_n(x) = \sum_{1}^{n} f_p(x)$$

converges to a differentiable sum function $s(x)$ in (a, b), show that the series $\sum\limits_{1}^{\infty} f'_n(x)$ converges in (a, b) to a function $\lambda(x)$ such that $\lambda(x) \leqslant s'(x)$.
(Show that $s(x+h) - s(x) \geqslant \sum\limits_{1}^{n} \{f_p(x+h) - f_p(s)\}$.)

(ii) Show that there is a sequence of integers p_n such that the series

$$\sum_{n=1}^{\infty} \{s(x) - s_{p_n}(x)\}$$

converges for each x in (a, b). Hence deduce that

$$s'_{p_n}(x) \to s'(x)$$

and
$$s'_n(x) \to s'(x)$$

as $n \to \infty$. (This is a mild form of a theorem due to Fubini. Note that $s(x) - s_{p_n}(x) \leqslant s(b) - s_{p_n}(b)$.)

6. If $\phi(x) = f(x) + g(x)$, prove that

$$D_+ f + D_+ g \leqslant D_+ \phi \leqslant D^+ \phi \leqslant D^+ f + D^+ g.$$

By considering the functions

$$f(x) = |x|, \qquad g(x) = x - |x|,$$

show that the signs $\leqslant$ cannot be replaced by $=$.

7. If $f(x)$ has a unique derivative $f'(x)$ prove that

$$D^+(f+g) = f'(x) + D^+ g.$$

4 Indicators

4.1. Introduction

We have it on the authority of Henri Poincaré (*Œuvres de Laguerre*, tome 1, Préface, p. x, Paris, 1898) that 'in the mathematical sciences a good notation has the same philosophical importance as a good classification in the natural sciences'. In the theory of sets of points there are many advantages in adopting the notation due to Charles de la Vallée-Poussin (1916), by which the whole of the theory is expressible in analytical form, rather than in the usual geometrical language. In particular, it is unnecessary to memorize formulae for the manipulation of special symbols for the intersection and complements of sets of points.

DEFINITION 4.1.1. The 'characteristic function', $\chi(x, E)$ of a set of points E in a space R is defined by the relations

$$\chi(x, E) = \begin{cases} 1 & (x \in E), \\ 0 & (x \notin E), \end{cases}$$

x being any point of R.

In view of the numerous meanings that have been given to the adjective 'characteristic', we shall follow the lead of modern books on probability theory and call the function $\chi(x, E)$ the 'indicator' of the set E. Thus the indicator of the whole space R is the function $f(x) \equiv 1$ and the indicator of the 'empty set' is $f(x) \equiv 0$.

When we are discussing the properties of some specified set E we may write $\chi(x)$ for $\chi(x, E)$ and often we may further abbreviate $\chi(x)$ to the single letter χ.

Just as we commonly speak of 'the point (x, y)', meaning the point with coordinates (x, y), so we shall speak of the 'set' or 'set of points $\chi(x, E)$', meaning the set of points E with indicator $\chi(x, E)$.

The fundamental property of indicators is given by

THEOREM 4.1.1. *The necessary and sufficient condition that a function $f(x)$, i.e. a mapping from the space S of the points x to the space R of the real numbers, should be an indicator is that*

$$\{f(x)\}^2 = f(x) \quad \text{for each } x.$$

Since the elements a of a Boolean algebra are characterized by the relation $a^2 = a$ it seems appropriate to give the theory of indicators the name of *Boolean analysis*. Thus point-set topology can be expressed in the form of Boolean analysis, and we shall proceed to summarize the relevant properties of sets of points in this form.

The first fundamental relation in point-set theory is that of 'inclusion'.

DEFINITION 4.1.2. A set α is 'included' in a set β, or 'covered' by a set β if each point of α is also a point of β.

THEOREM 4.1.2. *The necessary and sufficient condition that the set α should be 'covered' by the set β is that*

$$\alpha(x) = \alpha(x)\beta(x) \quad \text{for each } x.$$

For, if α is covered by β then, by Definition 4.1.2,

$$\alpha(x) = 1 \quad \text{implies that} \quad \beta(x) = 1,$$

whence $\qquad\qquad \alpha(x) \leqslant \beta(x) \quad \text{for each } x.$

Now $\qquad\qquad 0 \leqslant \alpha(x) \leqslant 1 \quad \text{and} \quad 0 \leqslant \beta(x) \leqslant 1,$

whence $\qquad\qquad 0 \leqslant \alpha(1-\beta) \leqslant \beta(1-\beta) = 0,$

and $\qquad\qquad\qquad \alpha(1-\beta) = 0,$

i.e. $\qquad\qquad\qquad \alpha(x) = \alpha(x)\beta(x).$

Conversely, if $\alpha = \alpha\beta$ then, either $\alpha = 0$ or $\beta = 1$ and in either case $\alpha \leqslant \beta$.

The second fundamental relation is that of 'disjunction'.

DEFINITION 4.1.3. A finite or enumerable collection of sets, $\alpha_1, \alpha_2, \ldots$, is said to be 'disjoint' if $\alpha_p \alpha_q = 0$ for $p \neq q$, i.e. if no two different sets have a point in common.

4.2. Boolean convergence

Since it is one of the distinguishing features of the Lebesgue theory of measure to consider enumerable collections of sets of points, we proceed at once to consider the conditions for the convergence of a sequence of indicators $\{\chi_n(x)\}$ $(n = 1, 2, ...)$.

In classical analysis the condition for the convergence of a sequence $\{s_n\}$ $(n = 1, 2, ...)$ to a limit s as n tends to infinity is that, if ϵ is any prescribed tolerance, then

$$|s_n - s| < \epsilon$$

for 'almost all n', i.e. for all values of n except a finite number, i.e. for all n greater than some threshhold $n(\epsilon)$ which usually depends on ϵ. But in a sequence of indicators, $\{\chi_n(x)\}$, each term can take two values only, viz. 0 and 1. Hence if $\chi_n(x)$ converges to a limit $\chi(x)$ the condition that

$$|\chi_n(x) - \chi(x)| < \epsilon$$

for almost all n implies that, for each value of x, the terms $\chi_n(x)$ are actually equal to the limit $\chi(x)$ for almost all n. Hence we have

THEOREM 4.2.1. *The necessary and sufficient condition for the convergence of a sequence of indicators $\{\chi_n(x)\}$ to a limit $\chi(x)$ as $n \to \infty$ is that, for each x,*

$$\chi_n(x) = \chi(x)$$

for almost all n.

COROLLARY. *The limit $\chi(x)$ of a convergent sequence of indicators is itself an indicator.*

For almost all n, and each value of x,

$$\chi^2 = \chi_n^2 = \chi_n = \chi.$$

Hence by Definition 4.1.1, χ is an indicator.

THEOREM 4.2.2. *If $\{\chi_n(x)\}$ is any enumerable sequence of indicators, then the product*

$$\pi_n(x) = \chi_1(x)\,\chi_2(x)\,\cdots\,\chi_n(x)$$

is also an indicator, and the sequence $\{\pi_n(x)\}$ is always convergent.

For, if $\chi_n(\xi) = 1$ for all n, then $\pi_n(\xi) = 1$ for all n and $\pi_n(\xi) \to 1$. But, if there is at least one integer p such that $\chi_p(\xi) = 0$, then $\pi_n(\xi) = 0$ if $n \geqslant p$, and $\pi_n(\xi) \to 0$. Hence $\pi_n(\xi)$ tends to a limit $\pi(x)$ which has the values 0 and 1 only, and is therefore an indicator.

Therefore we can frame

DEFINITION 4.2.1. The 'intersection' of a finite or enumerable collection of sets $\{\chi_n(x)\}$ ($n = 1, 2,...$) is the set

$$\pi(x) = \chi_1(x)\chi_2(x) \cdots \chi_n(x)$$

if the collection is finite, or the set

$$\pi(x) = \lim_{n\to\infty} \{\chi_1(x)\chi_2(x) \cdots \chi_n(x)\}$$

if the collection is enumerable.

THEOREM 4.2.3. *The intersection of* $\chi_1, \chi_2,...$ *consists of the points common to* $\chi_1, \chi_2,...$.

For $\pi(x) = 1$ if and only if $\chi_n(x) = 1$ for all n.

DEFINITION 4.2.2. The 'union' of a finite or enumerable collection of sets $\{\chi_m(x)\}$ ($m = 1, 2,...$) is the set

$$\sigma(x) = 1 - \{1-\chi_1(x)\}\{1-\chi_2(x)\} \cdots \{1-\chi_n(x)\}$$

if the collection is finite, or the set

$$\sigma(x) = 1 - \lim_{n\to\infty} \{1-\chi_1(x)\}\{1-\chi_2(x)\} \cdots \{1-\chi_n(x)\}$$

if the collection is enumerable.

THEOREM 4.2.4. *The union of the sets* $\chi_1, \chi_2,...$ *consists of the points which belong to at least one of these sets.*

For $\sigma(x) = 1$ only if one of the terms $\{1-\chi_p(x)\}$ is zero.

Although we need no special symbol, such as $\bigcap \chi_m(x)$, for the intersection of the sets $\chi_1, \chi_2,...$, it will be convenient to denote their union by $\bigcup \chi_n(x)$, or by $\bigcup_{n=1}^{\infty} \chi_n(x)$, or by $\alpha \cup \beta$, if there are only two sets α and β.

A special case of great importance is covered by

THEOREM 4.2.5. *If the collection of sets $\{\chi_n(x)\}$ is disjoint, then the series*

$$\sum_{n=1}^{\infty} \chi_n(x)$$

always converges to the indicator of the union of the sets.

For $\chi_p \chi_q = 0$ if $p \neq q$, whence

$$1-(1-\chi_1)(1-\chi_2)...(1-\chi_n) = \chi_1+\chi_2+...+\chi_n.$$

4.3. Open and closed sets

If we employ the language of Boolean analysis then the simplest and most natural way to define the classical concepts of 'open' and 'closed' sets is as follows.

DEFINITION 4.3.1. The set $o(x)$ is said to be 'open' if the indicator $o(x)$ is continuous at each point $x = \xi$ where $o(\xi) = 1$.

DEFINITION 4.3.2. The set $\kappa(x)$ is said to be 'closed' if the indicator $\kappa(x)$ is continuous at each point $x = \xi$ where $\kappa(\xi) = 0$.

THEOREM 4.3.1. *The open interval $a < x < b$ is an open set in the sense of Definition 4.3.1, and the closed interval $a \leqslant x \leqslant b$ is a closed set in the sense of Definition 4.3.2.*

In particular, an isolated point, $x = c$, is a closed set.

THEOREM 4.3.2. *If ξ is a point of the open set $o(x)$ then ξ is an interior point of this set.*

For there is a neighbourhood of ξ in which $o(x) = o(\xi) = 1$. Hence this neighbourhood of ξ belongs to the set $o(x)$, i.e. ξ is an interior point of $o(x)$.

THEOREM 4.3.3. *If $\kappa(x)$ is a closed set, then it is identical with its closure.*

Let $\xi_1, \xi_2,...$ be any convergent sequence of points belonging to $\kappa(x)$ with limit ξ. Then ξ belongs to the closure of $\kappa(x)$, and any neighbourhood of ξ contains an infinite number of the points $\{\xi_n\}$. But, if $\kappa(\xi) = 0$, there exists a neighbourhood of ξ for all points x of which $\kappa(x) = 0$ by Definition 4.3.2. Thus we have a contradiction and hence $\kappa(\xi) = 1$, i.e. the set $\kappa(x)$ contains the limit of any convergent sequence of points in $\kappa(x)$.

The preceding theorems show that the definitions of open and closed sets in terms of the continuity of their indicators are equivalent to the usual definitions. (But see Exercise 7 of § 4.60.)

DEFINITION 4.3.3. Two sets $\alpha(x)$ and $\beta(x)$ are said to be 'complementary' with respect to a set $\gamma(x)$ if $\alpha(x)$ and $\beta(x)$ are disjoint, and $\gamma(x)$ is their union.

THEOREM 4.3.4. *The necessary and sufficient condition that* $\alpha(x)$ *and* $\beta(x)$ *should be complementary with respect to* $\gamma(x)$ *is that*

$$\alpha(x)+\beta(x) = \gamma(x).$$

If α and β are disjoint their union is

$$1-(1-\alpha)(1-\beta) = \alpha+\beta.$$

Hence the condition is necessary.

It is also sufficient, for it implies that

$$0 = (\alpha+\beta-\gamma)(\alpha+\beta+\gamma) = \alpha+\beta-\gamma+2\alpha\beta = 2\alpha\beta,$$

whence α and β are disjoint. Therefore their union is

$$\alpha+\beta = \gamma.$$

THEOREM 4.3.5. *The complement* $o(x)$ *of a closed set* $\kappa(x)$ *with respect to an open set* $\Omega(x)$ *is itself open.*

For $\qquad\qquad o(x) = \Omega(x)-\kappa(x).$

Hence, if $o(\xi) = 1$, then $\Omega(\xi) = 1$ and $\kappa(\xi) = 0$.

Therefore $\Omega(x)$ and $\kappa(x)$ are continuous at ξ. Therefore $o(x)$ is continuous at ξ. Hence $o(x)$ is open.

THEOREM 4.3.6. *The complement* $\kappa(x)$ *of an open set* $o(x)$ *with respect to a closed set* $\Gamma(x)$ *is itself closed.*

For $\qquad\qquad \kappa(x) = \Gamma(x)-o(x).$

Hence, if $\kappa(\xi) = 0$, then

either (i) $\Gamma(\xi) = 1 = o(\xi)$,

or (ii) $\Gamma(\xi) = 0 = o(\xi)$.

If $o(\xi) = 1$ then there is a neighbourhood of ξ in which $o(x) = 1$ and in which

$$\Gamma(x) = \kappa(x) + o(x) \geqslant o(x)$$

whence $\Gamma(x) = 1$ and $\Gamma(x)$ is continuous. Hence $\kappa(x)$ is continuous at ξ.

If $\Gamma(\xi) = 0$ then there is a neighbourhood of ξ in which $\Gamma(x) = 0$ and in which

$$\kappa(x) = \Gamma(x) - o(x) \leqslant \Gamma(x),$$

whence $\kappa(x) = 0$. Thus $\kappa(x)$ is continuous at ξ.

Hence $\kappa(x)$ is continuous at any point ξ at which $\kappa(\xi) = 0$. Therefore $\kappa(x)$ is closed.

THEOREM 4.3.7. *The intersection of a finite or enumerable collection of closed sets $\{\kappa_p\}$ is closed.*

Let

$$\alpha_n(x) = \kappa_1(x)\kappa_2(x) \dots \kappa_n(x),$$

and

$$\alpha_n(\xi) = 0.$$

Then there exists an integer p such that

$$1 \leqslant p \leqslant n$$

and

$$\kappa_p(\xi) = 0.$$

Hence there is a neighbourhood of ξ in which $\kappa_p(x) = 0$, and therefore $\alpha_n(x) = 0$. Thus $\alpha_n(x)$ is closed.

Now let

$$\alpha(x) = \lim_{n \to \infty} \alpha_n(x),$$

and

$$\alpha(\xi) = 0.$$

Then, by Theorem 4.2.1, there is an integer p such that $\alpha_p(\xi) = 0$. Hence there is a neighbourhood of ξ in which $\alpha_p(x) = 0$ and therefore $\alpha(x) = 0$. Therefore $\alpha(x)$ is closed.

THEOREM 4.3.8. *The union of a finite or enumerable collection of open sets is open.*

This follows as the 'dual' theorem of 4.3.7 by using the principles of complementarity (4.3.5 and 4.3.6).

4.4. Covering theorems

THEOREM 4.4.1. *The union of enumerable non-disjoint sets* $\{\alpha_n\}$ *is also the union of the enumerable disjoint sets* $\{\beta_n\}$ *defined as*

$$\beta_1 = \alpha_1,$$
$$\beta_n = \alpha_n(1-\alpha_1)(1-\alpha_2)\dots(1-\alpha_{n-1}) \quad (n > 1).$$

Since
$$\alpha_n^2 = \alpha_n,$$

and
$$(1-\alpha_n)^2 = (1-\alpha_n),$$

it follows that $\beta_n^2 = \beta_n$.

Hence the functions $\{\beta_n\}$ are indicators.

Also, if $p < q$,

$$\beta_p\beta_q = \alpha_p\,\alpha_q(1-\alpha_1)(1-\alpha_2)\dots(1-\alpha_{q-1}).$$

The right-hand side contains the factor

$$\alpha_p(1-\alpha_p) = 0,$$

whence
$$\beta_p\beta_q = 0,$$

so that the sets $\{\beta_p\}$ are disjoint.

Finally, we can prove by induction that

$$\beta_1+\beta_2+\dots+\beta_n = 1-(1-\alpha_1)(1-\alpha_2)\dots(1-\alpha_n).$$

For
$$\beta_1 = 1-(1-\alpha_1)$$

and
$$1-(1-\alpha_1)(1-\alpha_2)\dots(1-\alpha_n)+\beta_{n+1}$$
$$= 1+(1-\alpha_1)(1-\alpha_2)\dots(\alpha_{n+1}-1).$$

Hence
$$\sum_{n=1}^{\infty}\beta_n = 1-\prod_{n=1}^{\infty}(1-\alpha_n),$$

the infinite series and product converging by Theorem 4.2.2.

COROLLARY. *The union of enumerable non-disjoint intervals is also the union of certain enumerable disjoint intervals.*

For, if $\alpha_1, \alpha_2,\dots$ are intervals, then each β_n is a finite collection of disjoint intervals.

THEOREM 4.4.2 (the Heine–Borel theorem). *If* $\alpha(x)$ *is a compact set of points E covered by an enumerable collection C of open sets* $\{\alpha_n(x)\}$ $(n = 1, 2,...)$ *then* E *is also covered by a finite number of sets in the collection* C.

Let
$$\phi_n(x) = \sum_{k=1}^{n} \alpha_k(x).$$

Then the possible values of the function $\phi_n(x)$ are $0, 1, 2,..., n$. Hence $\phi_n(x)$ has a greatest lower bound λ_n which is certainly attained at some point ξ_n, and

$$\phi_n(x) \geqslant \phi_n(\xi_n) = \lambda_n \quad \text{for all } x \text{ in } E.$$

Since E is compact, the sequence of points $\{\xi_n\}$ has at least one point of accumulation ξ in E, and moreover, for some integer q, $\alpha_q(\xi) = 1$.

But $\alpha_q(x)$ is continuous at ξ since α_q is open. Therefore, for some integer $p \geqslant q$,
$$\alpha_q(\xi_p) = 1.$$

Hence
$$\phi_p(\xi_p) \geqslant \alpha_q(\xi_p) = 1.$$

Thus
$$\phi_p(x) \geqslant \phi_p(\xi_p) \geqslant 1$$

and E is covered by the finite number of open sets, $\alpha_1, \alpha_2,..., \alpha_p$.

COROLLARY. *The Heine–Borel theorem clearly applies if* E *is a bounded, closed set in* d-*dimensional Euclidean space.*

4.5. The indicator of a function

We have already indicated in § 2.4 the importance of the functions $\alpha_p(x)$ associated with a function $f(x)$ by the relations

$$\alpha_p(x) = \begin{cases} 1 & \text{if } t_p < f(x) \leqslant t_{p+1}, \\ 0 & \text{if } f(x) \leqslant t_p \text{ or } t_{p+1} < f(x). \end{cases}$$

In general, if x is a point of a space S and $f(x)$ any mapping from S to the real numbers R we have the definition

DEFINITION 4.5.1. *The indicator of the function* $f(x)$, *viz.* $\alpha(x, t, f)$ *or* $\alpha(x, t)$ *is defined as the indicator of the set of points* x *at which*
$$f(x) > t.$$

Hence
$$\alpha_p(x) = \alpha(x, t_p) - \alpha(x, t_{p+1}).$$

THEOREM 4.5.1. *The indicator of the function* $f(x)$ *is a non-increasing function of* t.

For, if $s < t$ and $f(x) > t$, then $f(x) > s$, i.e. $\alpha(x,t,f) = 1$ implies that $\alpha(x,s,f) = 1$.

But, if $f(x) \leqslant t$, then we may have either $f(x) > s$ or $f(x) \leqslant s$, i.e. $\alpha(x,t,f) = 0$ implies that $\alpha(x,s,f) = 1$ or 0. Therefore
$$\alpha(x,t,f) \leqslant \alpha(x,s,f).$$

4.6. Exercises

1. If the indicators α, β and $\alpha\beta$ are linearly dependent, i.e. if there exists a linear relation
$$a\alpha + b\beta + c\alpha\beta = 0$$
such that the real numbers a, b, c are not all zero, prove that either $\alpha = \alpha\beta$ or $\beta = \alpha\beta$.

2. If α, β, $\alpha\beta$, and γ are each indicators and
$$\gamma = a\alpha + b\beta + c\alpha\beta,$$
where a, b, c are real numbers, prove that

either $\qquad\qquad \gamma = \alpha + \beta - \alpha\beta,$

or $\qquad\qquad \gamma = \alpha + \beta - 2\alpha\beta,$

or $\qquad\qquad \gamma = \alpha - \alpha\beta,$

or $\qquad\qquad \gamma = \beta - \alpha\beta.$

3. The 'differences' of two sets α and β are defined as
$$\alpha \backslash \beta = \alpha - \alpha\beta$$
and $\qquad\qquad \beta \backslash \alpha = \beta - \alpha\beta$
while the 'symmetric difference' is defined as
$$\alpha \Delta \beta = \alpha + \beta - 2\alpha\beta.$$
Give geometric definitions of these sets in terms of the relations of inclusion and disjunction.

4. If ξ is a point covered by the union of the sets $\{\alpha_n(x)\}$ of Theorem 4.4.1, prove directly that there is an integer p such that
$$\alpha_n(\xi) = 0 \quad \text{if } n < p,$$
$$\alpha_p(\xi) = 1,$$
$$\beta_n(\xi) = 0 \quad \text{if } n \neq p,$$
$$\beta_p(\xi) = 1.$$

5. If $\{\alpha_n(x)\}$ is a decreasing sequence of non-empty, bounded closed sets in d-dimensional Euclidean space, i.e. if
$$\alpha_1 > \alpha_2 > \dots > \alpha_n > \alpha_{n+1} > \dots,$$
prove that the intersection of all the α_n is not empty.

6. Show that the indicator of the rational points in $[0, 1]$ can be expressed in the form
$$\chi(x) = \lim_{m \to \infty} \left\{ \lim_{n \to \infty} (\cos m!\,\pi x)^{2n} \right\}.$$

5   Differentiation of monotone functions

5.1. Introduction

In Chapter 1 in order to provide a motivation for the search for the most general concept of integration we introduced the ideas of 'tame' and 'wild' functions. The tame functions of elementary calculus are the bounded continuous functions with finite derivatives everywhere. Then there are functions such as

$$\text{sgn}\, x, \quad |x|, \quad [x],$$

which are tame everywhere except at one point (in these cases the origin). From these we can construct functions such as

$$\sum_{p=1}^{q} |x-p/q| \quad (p, q \text{ integers}), \qquad \sum_{p=1}^{q} \text{sgn}(x-p/q),$$

which are tame everywhere except at a finite number of points; and functions such as

$$\sum_{n=1}^{\infty} \frac{|x-r_n|}{3^n},$$

which is tame everywhere except at the rational points $x = r_1, r_2,\ldots$.

Thus it appears that there are degrees of wildness and that one way of quantifying the wildness of a function is by giving some general specification of the points at which it ceases to be tame. Such a criterion of wildness should have some practical utility and be related to the general concept of integration. Thus if the 'wild' points of a wild function could be neglected in constructing its integral, the function could be described as 'wild but harmless'.

For example all physicists would regard $|x|$ as a primitive of sgn x disregarding the discontinuity in sgn x at the origin. Similarly any finite number of discontinuities in an otherwise continuous integral are commonly ignored. But how far can we go in neglecting wild points? What is the largest collection of wild points that can be safely ignored in integration?

The answer to this question is furnished by Lebesgue's theory of sets of points of zero measure.

5.2. Sets of points of measure zero

The length of an interval I will be denoted by $|I|$ and the area of a rectangle R by $|R|$.

DEFINITION 5.2.1. A set of points E on the real axis is said to have zero one-dimensional measure if, to each positive number ϵ there corresponds an enumerable collection of intervals $\{I_n\}$ ($n = 1, 2,...$), which cover the set E and whose total length

$$I = \sum_{n=1}^{\infty} |I_n|$$

does not exceed ϵ.

We note that the intervals may possibly be overlapping, and that it is immaterial whether the intervals are open, closed, or half-open. Also, since the terms $|I_1|$, $|I_2|$,... of the series are each positive, the sum I is independent of the order of enumeration.

DEFINITION 5.2.2. A set of points E in the (x, y)-plane is said to have zero two-dimensional measure, if to each positive number ϵ there corresponds an enumerable collection of rectangles $\{R_n\}$ (with edges parallel to the lines $x = 0$ or $y = 0$) which cover the set E and whose total area

$$R = \sum_{n=1}^{\infty} |R_n|$$

does not exceed ϵ.

As before, the rectangles may possibly be overlapping and may or may not include their edges or vertices.

When there is no danger of confusion we shall use the shorter description 'sets of zero measure', or 'null sets' for sets with zero one-dimensional measure.

When a function $f(x)$ possesses a property P for each point of an interval (a, b) except for the points of a set of measure zero it is customary to say that '$f(x)$ possesses the property P almost everywhere in (a, b)' or to write '$f(x)$ possesses the property P p.p. in (a, b)' (p.p. being the abbreviation for *presque partout*).

The essential point in these definitions, due effectively to Lebesgue, is the use of an *enumerable* collection of intervals (or rectangles) to cover the set of points E.

It is obvious that, in one dimension, a single point, or a finite number of points has zero measure, and we shall prove that the same is true for an enumerable set of points.

THEOREM 5.2.1. *Any enumerable set of points* $E = \{x_1, x_2, ...\}$ *has zero measure.*

For E is covered by the enumerable collection of intervals

$$I_n = \left\{ x, \ x_n - \frac{\epsilon}{2^{n+1}} < x < x_n + \frac{\epsilon}{2^{n+1}} \right\}$$

whose total length is

$$I = \sum_{n=1}^{\infty} |I_n| = \sum_{n=1}^{\infty} \frac{\epsilon}{2^n} = \epsilon.$$

COROLLARY. *The set of rational points, i.e. the points whose coordinates* $x_1, x_2, ...$ *are each a rational number, is enumerable and therefore has zero measure.*

THEOREM 5.2.2. *If* $E_1, E_2, ...$ *is an enumerable collection of sets, each of measure zero, then their union* E *is also of measure zero.*

For the set E_k can be covered by an enumerable collection of intervals (a_{kn}, b_{kn}) $(n = 1, 2, ...)$ of total length less than $\epsilon/2^k$, where ϵ is any arbitrary tolerance. Hence the union of $E_1, E_2, ...$ can be covered by an enumerable collection of intervals of total length less than ϵ.

The question naturally arises whether a set of points of zero measure is necessarily enumerable. The answer is in the negative as is shown by the example of the 'Cantor set' in the next section.

5.3. The Cantor set of points

To construct the Cantor set of points we proceed as follows.

From the unit interval, $0 \leqslant x \leqslant 1$, we remove in succession

(i) E_1, the middle third, $\frac{1}{3} < x < \frac{2}{3}$,

(ii) E_2, the middle thirds of the remaining intervals, viz.

$$\frac{1}{9} < x < \frac{2}{9}, \qquad \frac{7}{9} < x < \frac{8}{9},$$

(iii) E_3, the middle thirds of the remaining intervals, viz.

$$\frac{1}{27} < x < \frac{2}{27}, \quad \frac{7}{27} < x < \frac{8}{27}, \quad \frac{19}{27} < x < \frac{20}{27}, \quad \frac{25}{27} < x < \frac{26}{27},$$

and continue this process indefinitely. The lengths of the intervals removed are

$$|E_1| = \tfrac{1}{3}, \qquad |E_2| = \frac{2}{3^2}, \qquad |E_3| = \frac{4}{3^3},$$

and, in general, $\qquad |E_n| = \tfrac{1}{3}(\tfrac{2}{3})^{n-1}.$

Thus, at the nth stage, we have removed intervals of total length

$$\sum_{k=1}^{n} |E_k| = 1 - (\tfrac{2}{3})^n.$$

Now consider the points that remain after $E_1, E_2,...$ have been removed. These form a set C. C is certainly covered by the points that remain after $E_1, E_2,..., E_n$ have been removed. The removal of this finite number of intervals leaves a set of points C_n which consists of a finite number of intervals of total length $(2/3)^n$. Since n can be chosen to make $(2/3)^n$ less than any assigned tolerance $\epsilon > 0$, the set C has zero measure.

The set C, first constructed by Cantor, is not, however, enumerable. To prove this we observe that any number x in the interval $0 \leqslant x \leqslant 1$ can be expressed in the form

$$x = \sum_{n=1}^{\infty} \frac{a_n}{3^n},$$

where each numerator a_n is either 0, 1, or 2. The points of E_1 can be expressed as

$$x = \sum_{n=2}^{\infty} \frac{a_n}{3^n} + \frac{1}{3};$$

the points of E_2 as

$$x = \sum_{n=3}^{\infty} \frac{a_n}{3^n} + \tfrac{1}{9} + (0 \text{ or } \tfrac{2}{3});$$

the points of E_3 as

$$x = \sum_{n=4}^{\infty} \frac{a_n}{3^n} + \tfrac{1}{27} + (0 \text{ or } \tfrac{2}{3} \text{ or } \tfrac{2}{9} \text{ or } \tfrac{2}{3} + \tfrac{2}{9}).$$

In general, the points of E_n are those for which x can be expressed in the form

$$\sum_{k=1}^{n-1} \frac{a_k}{3^k} + \frac{1}{3^n} + \sum_{k=n+1}^{\infty} \frac{a_k}{3^k},$$

where $a_k = 0$, 1, or 2, i.e.

$$x = \sum_{n=1}^{\infty} \frac{a_k}{3} \quad \text{and} \quad a_n = 1.$$

Hence the points of the Cantor set C, i.e. the points of the interval $(0, 1)$ that remain after the removal of $E_1, E_2,...$, can be represented only in the form

$$x = \sum_{k=1}^{\infty} \frac{a_k}{3^k},$$

where each a_k is either 0 or 2.

If possible let these points be enumerable. Then they will form a sequence $\{x_n\}$, with $n = 1, 2,...$, and x_n expressible in the form

$$x_n = \sum_{k=1}^{\infty} \frac{a_{n,k}}{3^k},$$

where each $a_{n,k}$ is either 0 or 2.

Now consider the point

$$\xi = \sum_{k=1}^{\infty} \frac{C_k}{3^k}$$

where $C_k = 2 - a_{k,k}$.

This point lies in the interval $(0, 1)$ and belongs to the Cantor set since $C_k = 0$ or 2. But C_k is always different from $a_{k,k}$. Hence ξ is different from each number x_n, for $n = 1, 2,...$, i.e. the number

ξ that belongs to C is not included in the given enumeration of the points of C. We have thus arrived at a contradiction, which proves that the Cantor set C is not enumerable.

5.4. Average metric density

A simple and useful criterion to prove that a set of points E has zero measure (Boas 1960, p. 64), is conveniently expressed in terms of the concept of the 'average metric density' of the set E in an interval I.

DEFINITION 5.4.1. If the subset of E that lies in the interval I, of length $|I|$, can be covered by an enumerable collection of intervals of total length not greater than $\rho|I|$, then we say that the 'average metric density' of the set E in the interval I is not greater than ρ.

Clearly $0 \leqslant \rho \leqslant 1$.

THEOREM 5.4.1. *If the average metric density of a bounded set E is not greater than a number ρ less than unity, for all intervals I, then E is a set of measure zero.*

Consider the subset of E in an interval (a, b).

This subset can be covered by an enumerable collection of intervals (a_k, b_k) $(k = 1, 2, ...)$ of total length not greater than $\rho(b-a)$.

Now the subset of E in (a_k, b_k) can be covered by an enumerable collection of intervals (a_{kl}, b_{kl}) $(l = 1, 2, ...)$ of total length not greater than $\rho(a_k - b_k)$.

Hence the subset of E in (a, b) can be covered by an enumerable collection of intervals (a_{kl}, b_{kl}) $(k, l = 1, 2, ...)$ of total length not greater than

$$\rho \sum_{k=1}^{\infty} (a_k - b_k) \leqslant \rho^2 (b-a).$$

By mathematical induction it follows that the subset of E in (a, b) can be covered by an enumerable collection of intervals of total length not greater than $\rho^n(b-a)$, where n is any positive integer. Since n is arbitrary, the subset of E in (a, b) has zero measure.

5.5. The problem of differentiation

We are now in a position to enunciate the central result of the Lebesgue theory of differentiation as follows.

If the function $f(x)$ is continuous and non-decreasing in the interval (a, b), then $f(x)$ possesses a derivative $f'(x)$ at all points of this interval, with the exception of a set of points Z of zero measure.

This remarkable result, which finds so many applications, not only in the theory of integration but also in the whole of analysis and differential geometry, was first discovered by Lebesgue (1904, p. 128). The line of argument that we shall follow is that given by F. Riesz (1953, pp. 6–7) as simplified by R. P. Boas (1960, p. 134).

The main strategy is to show that the following chain of inequalities hold almost everywhere in (a, b), viz.

$$0 \leqslant D^+f(x) \leqslant D_-f(x) \leqslant D^-f(x) \leqslant D_+f(x) \leqslant D^+f(x) < \infty.$$

It then follows at once that the four Dini derivatives of $f(x)$ are finite and equal almost everywhere, i.e. $f(x)$ possesses a finite derivative almost everywhere in (a, b).

To establish these inequalities it is sufficient to prove that, if $f(x)$ is any continuous, non-decreasing function, then

$$D^+f(x) < \infty \quad \text{p.p.}$$

and
$$D^+f(x) \leqslant D_-f(x) \quad \text{p.p.} \tag{I}$$

For, if $y = -x$

and
$$g(y) = -f(x),$$

then
$$g(y+h) = -f(-y-h) = -f(x-h),$$

and
$$\frac{g(y+h)-g(y)}{h} = \frac{f(x-h)-f(x)}{-h}.$$

Hence
$$D^+g(y) = D^-f(x)$$

and
$$D_-g(y) = D_+f(x).$$

But $g(y+h)-g(y) = f(x)-f(x-h)$, so that the function $g(y)$, like $f(x)$, is continuous and non-decreasing in y.

Therefore (I) implies that

$$D^+g(x) \leqslant D_-g(x) \quad \text{p.p.}$$

whence $\qquad\qquad D^-f(x) \leqslant D_+f(x) \quad \text{p.p.} \qquad\qquad\text{(II)}$

Now by the very definition of the Dini derivatives

$$D_-f(x) \leqslant D^-f(x)$$

and $\qquad\qquad D_+f(x) \leqslant D^+f(x). \qquad\qquad\text{(III)}$

Hence, by combining the inequalities (I), (II), and (III), we find the desired result:

$$D^+f(x) \leqslant D_-f(x) \quad \text{p.p.} \qquad\qquad\text{(I)}$$

$$\leqslant D^-f(x) \qquad\qquad\qquad\text{(III)}$$

$$\leqslant D_+f(x) \quad \text{p.p.} \qquad\qquad\text{(II)}$$

$$\leqslant D^+f(x). \qquad\qquad\qquad\text{(III)}$$

The problem is thus resolved into the proof of the two inequalities
$$D^+f(x) < \infty \quad \text{p.p.}$$

and $\qquad\qquad D^+f(x) \leqslant D_-f(x) \quad \text{p.p.}$

for any continuous, non-decreasing function $f(x)$.

5.6. The 'rising sun' lemma

Various methods have been devised to construct enumerable collections of intervals covering the points at which $D^+f(x) < \infty$ or $D^+f(x) \leqslant D_-f(x)$, such as those invented by Vitali (Burkill 1953, p. 46), de la Vallée-Poussin (1916), and Rajchman and Saks (Titchmarsh, p. 358). But the method devised by F. Riesz has the advantage of a simple geometric interpretation and we shall therefore adopt it here, following the vivid account given by R. P. Boas (1960, p. 134).

The geometrical significance of the 'rising sun' lemma is easily grasped if we regard the graph, $y = f(x)$, of a continuous function $f(x)$, as the profile of a series of parallel ridges (parallel to the z-axis !) of a mountain range, illuminated by the horizontal rays of the rising sun, at infinity on the x-axis (Fig. 2).

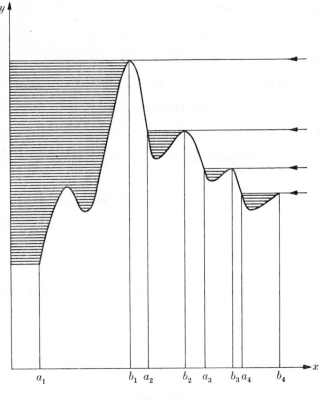

Fɪɢ. 2

Some points on the ridges are in the sunshine and some are in the shadows cast by ridges on their right. The points in the shadows occupy a number of hollows, such as

$$x' < x < x''$$

in which $\qquad\qquad f(x) < f(x'')$

whence $\qquad\qquad f(x') \leqslant f(x'').$

In order to give precision and rigour to these geometrical intuitions we frame the following definition and theorem.

DᴇFɪɴɪᴛɪᴏɴ 5.6.1. If $f(x)$ is continuous in the interval $[a, b]$, a point x is said to be shaded, or dominated, by a point ξ if

$$a \leqslant x < \xi \leqslant b \quad \text{and} \quad f(x) < f(\xi).$$

THEOREM 5.6.1 (the 'rising sun' lemma). *The dominated points form an enumerable collection of disjoint open intervals* (a_k, b_k) $(k = 1, 2,...)$ *such that if*

$$a_k < x < b_k,$$

then
$$f(x) < f(b_k)$$

and
$$f(a_k) \leqslant f(b_k).$$

The proof divides in four parts.

(1) Let $x = s$ be any dominated point. Since $f(x)$ is continuous, it attains a maximum $M(s)$ at some point ξ in the interval $[s, b]$; and $M(s) = f(\xi) > f(s)$. There may be several points ξ at which $f(x)$ attains the same maximum $M(s)$. Let s'' be the least upper bound of the points at which $f(\xi) = M(s)$. Then

$$f(x) < f(s'') \quad \text{if } s < x < s''.$$

(2) Since $f(x)$ is continuous, there is an interval $\sigma < x < s$ in which $f(x) \leqslant f(s'')$. Let s' be the greatest lower bound of the points σ at the lower extremity of these intervals. Then $s' < s$ and

$$f(x) \leqslant f(s''),$$

if
$$s' < x < s''.$$

Thus any dominated point $x = s$ lies in the interior of a unique 'shaded interval' (s', s'') such that, if $s' < x < s''$, then

$$f(x) < f(s'');$$

whence
$$f(s') \leqslant f(s'').$$

(3) The shaded intervals are disjoint.

For, if (s', s'') and (t', t'') are two overlapping shaded intervals, they have a common point x, and by (1),

$$f(x) < f(s''), \quad f(x) < f(t'').$$

Hence x is a dominated point. Therefore, by (1) and (2), x lies in a unique shaded interval. Thus the intervals (s', s'') and (t', t'') are identical.

Finally, to sum up, the dominated points s lie in a collection of disjoint open intervals, such as (s', s''), for which

$$f(s') \leqslant f(s'').$$

(4) The shaded intervals are enumerable.

For, if n is any positive integer, the total number of the shaded intervals in (a, b) with lengths greater than $(b-a)/2^n$ is not greater than 2^n, since the shaded intervals are disjoint. Hence the total number of shaded intervals in (a, b) with lengths lying between $(b-a)/2^{n+1}$ and $(b-a)/2^n$ is not greater than 2^{n+1} and therefore is finite. Thus the whole set of shaded intervals is enumerable.

There is a companion theorem, which we may call the 'setting sun lemma' and which is obtained by changing the definition of a dominated point as follows.

DEFINITION 5.6.2. If $f(x)$ is continuous in $[a, b]$, a point x is said to be dominated by a point if, for some ξ,

$$a \leqslant \xi < x \leqslant b \quad \text{and} \quad f(x) < f(\xi).$$

We can easily deduce the following theorem by the same methods as those employed for Theorem 3.6.1, putting $g(x) = f(-x)$.

THEOREM 5.6.2 (the 'setting sun' lemma). *The dominated points form an enumerable collection of disjoint open intervals (a_k, b_k) $(k = 1, 2, ...)$ such that if*

$$a_k < x < b_k,$$

then

$$f(x) < f(a_k)$$

and

$$f(b_k) \leqslant f(a_k).$$

5.7. The differentiation of monotone functions

We can now apply the covering theorems of the preceding section to examine the differentiability of monotone functions.

THEOREM 5.7.1. *If $\phi(x)$ is continuous and non-decreasing in $[a, b]$, if μ is any positive number, and E_μ the set of points in $[a, b]$ at which $D^+\phi(x) > \mu$, then E_μ is covered by an enumerable disjoint set of intervals P, (α_k, β_k) $(k = 1, 2, ...)$, such that*

$$g(P) \equiv \sum_{k=1}^{\infty} (\beta_k - \alpha_k) \leqslant \frac{\phi(b) - \phi(a)}{\mu}.$$

Let

$$f(x) = \phi(x) - \mu x.$$

Then $f(x)$ is continuous in $[a, b]$.

If $x \in E_\mu$, there exists a point ξ in (a, b) such that

$$x < \xi,$$

$$\frac{\phi(\xi) - \phi(x)}{\xi - x} > \mu,$$

i.e. $$f(\xi) > f(x),$$

i.e. x is dominated by ξ.

Therefore, by Theorem 3.6.1, the points E lie in an enumerable set of intervals, P, (α_k, β_k) $(k = 1, 2, \ldots)$, such that

$$f(\beta_k) \geqslant f(\alpha_k) \quad \text{for } k = 1, 2, \ldots .$$

Hence $$\phi(\beta_k) - \phi(\alpha_k) \geqslant \mu(\beta_k - \alpha_k).$$

Therefore $$\mu g(P) \leqslant \sum_{k=1}^{\infty} \{\phi(\beta_k) - \phi(\alpha_k)\}.$$

If we consider only a finite number of terms on the right-hand side, it is clear that, since $\phi(x)$ is non-decreasing, therefore

$$\sum_{k=1}^{n} \{\phi(\beta_k) - \phi(\alpha_k)\} \leqslant \phi(b) - \phi(a).$$

Since this is true for all n, therefore

$$\mu g(P) \leqslant \phi(b) - \phi(a).$$

THEOREM 5.7.2. *If $\phi(x)$ is continuous and non-decreasing in $[a, b]$, λ any positive number, and F_λ the set of points in $[a, b]$ at which $D_-(x) < \lambda$, then F_λ is covered by an enumerable set of intervals Q, (α_k, β_k) $(k = 1, 2, \ldots)$, such that*

$$\sum_{k=1}^{\infty} [\phi(\beta_k) - \phi(\alpha_k)] \leqslant \lambda(b - a).$$

Let $$f(x) = \phi(x) - \lambda x.$$

Then $f(x)$ is continuous in $[a, b]$. If $x \in F_\lambda$, there exists a point η in (a, b) such that
$$\eta < x$$

and $$\frac{\phi(\eta) - \phi(x)}{\eta - x} < \lambda,$$

i.e. $$f(\eta) > f(x).$$

Hence F_λ is covered by the set of points x at which $\eta < x$ implies that $f(x) < f(\eta)$.

Then, by the 'setting sun' lemma (5.6.2), these latter points form an enumerable collection of open intervals

$$(\alpha_k, \beta_k) \quad (k = 1, 2, ...)$$

such that $f(\alpha_k) \geqslant f(\beta_k).$

Hence, $\phi(\beta_k) - \phi(\alpha_k) \leqslant \lambda(\beta_k - \alpha_k).$

Therefore $\sum_{k=1}^{\infty} \{\phi(\beta_k) - \phi(\alpha_k)\} \leqslant \lambda(b-a).$

THEOREM 5.7.3. *If $\phi(x)$ is continuous and non-decreasing in a bounded interval $[a, b]$ then the set of points in $[a, b]$ at which*

$$D^+\phi(x) = +\infty$$

has measure zero.

For this set of points E is covered by an enumerable disjoint set of intervals P of total length $g(P)$ less than

$$\frac{\phi(b) - \phi(a)}{\mu}$$

by Theorem 5.7.1. Since this is true for all $\mu > 0$ it follows that E has measure zero.

THEOREM 5.7.4. *If $\phi(x)$ is continuous and non-decreasing in a bounded interval $[a, b]$, then the set of points in $[a, b]$ at which*

$$D_-\phi(x) < \lambda < \mu < D^+\phi(x)$$

has measure zero.

Apply Theorem 5.7.1 to a typical interval (α_k, β_k) of Theorem 5.7.2.

The set of points at which $D_-\phi(x) < \lambda$ is covered by the intervals (α_k, β_k). By Theorem 5.7.1, the set of points in $[\alpha_k, \beta_k]$ at which $D^+\phi(x) > \mu$ is covered by a disjoint collection of intervals of total length less than

$$\{\phi(\beta_k) - \phi(\alpha_k)\}/\mu.$$

Hence the set of points in $[a, b]$ at which

$$D_-\phi(x) < \lambda < \mu < D^+\phi(x)$$

is covered by a disjoint collection of intervals of total length less than $\mu^{-1} \sum_{k=1}^{\infty} \{\phi(\beta_k) - \phi(\alpha_k)\}$, and by Theorem 5.7.2 this is less than

$$\lambda(b-a)/\mu.$$

Therefore the average metric density of the points in (a, b) at which $D_-\phi(x) < \lambda < \mu < D^+\phi(x)$ is less than λ/μ which is less than unity.

Since this is true for any sub-interval (a, b), it follows from Theorem 3.4.1 that the set of points at which

$$D_-\phi(x) < \lambda < \mu < D^+\phi(x)$$

in a bounded interval is of measure zero.

THEOREM 5.7.5. *If $\phi(x)$ is continuous and non-decreasing in a bounded interval $[a, b]$ then $\phi(x)$ has a derivative $\phi'(x)$ almost everywhere in $[a, b]$.*

Take λ and μ to be rational numbers such that $\lambda < \mu$. Then by Theorem 5.7.4 the set of points $E_{\lambda,\mu}$ in $[a, b]$ at which

$$D_-\phi(x) < \lambda < \mu < D^+\phi(x)$$

is of measure zero. But the set of points $E_{\lambda,\mu}$ in $[a, b]$ at which $D_-\phi(x) < D^+\phi(x)$ is the union of the sets $E_{\lambda,\mu}$. Since λ and μ are rational these sets are enumerable. Hence, by Theorem 5.7.2, the set E has measure zero.

Therefore, almost everywhere,

$$
\begin{aligned}
0 \leqslant D^+\phi(x) &\leqslant D_-\phi(x) \\
&\leqslant D^-\phi(x) \quad \text{(III in § 5.5)} \\
&\leqslant D_+\phi(x) \quad \text{(II in § 5.5)} \\
&\leqslant D^+\phi(x) \quad \text{(III in § 5.5)} \\
&< \infty \qquad\quad \text{(Theorem 5.7.3)}.
\end{aligned}
$$

Hence the four Dini derivatives are equal and finite almost everywhere in (a, b), i.e. $\phi(x)$ is differentiable almost everywhere in (a, b).

5.8. The differentiation of series of monotone functions

Denjoy's theorem (§ 3.7, Exercise 4) on the integration of a monotone *sequence* of monotone functions has a companion theorem due to Fubini (Riesz and Sz-Nagy 1953, p. 12), on the differentiation of a convergent *series* of monotone functions.

THEOREM 5.8.1. *If $f_n(x)$ is a continuous, non-decreasing function of x in $[a, b]$ for each value of n $(1, 2,...)$ and if the series*

$$\sum_{n=1}^{\infty} f_n(x)$$

converges to a sum function $s(x)$ at each point of $[a, b]$, then almost everywhere in $[a, b]$, the series of derivatives

$$\sum_{n=1}^{\infty} f'_n(x)$$

exists and converges to the derivative $s'(x)$.

Apart from a set of values of x of measure zero the non-decreasing functions $f_n(x)$, and the partial sums

$$s_n(x) = \sum_{k=1}^{n} f_k(x),$$

together with the sum function $s(x)$, have finite, non-negative derivatives in (a, b). To study the convergence of the sequence

$$s'_n(x) = \sum_{k=1}^{n} f'_k(x)$$

we note that

$$s(x+h)-s(x) = \sum_{k=1}^{\infty} \{f_k(x+h)-f_k(x)\} \geqslant \sum_{k=1}^{n} \{f_k(x+h)-f_k(x)\}$$

whence $\qquad s'(x) \geqslant s'_n(x) \quad$ p.p. in (a, b)

The functions $f'_k(x)$ are non-negative, whence $\{s'_n(x)\}$ is a non-decreasing sequence. We have just shown that it is bounded by $s'(x)$. Hence the sequence $\{s'_n(x)\}$ is convergent almost everywhere in $[a, b]$.

To show that its limit is $s(x)$, we select a sub-sequence $s_{n(k)}(x)$, such that

$$0 \leqslant s(b)-s_{n(k)}(b) \leqslant 1/2^k \quad (k = 1, 2,...).$$

Then $\qquad s(x)-s_{n(k)}(x) = \sum_{p=n(k)+1}^{\infty} f_p(x),$

whence $s(x)-s_{n(k)}(x)$ is a non-decreasing function of x. Hence

$$0 \leqslant s(x)-s_{n(k)}(x) \leqslant s(b)-s_{n(k)}(b) \leqslant 1/2^k.$$

Therefore the series $\qquad \sum_{k=1}^{\infty} \{s(x)-s_{n(k)}(x)\}$

is a convergent series of non-decreasing functions.

We can apply the result established above for the similar series $\sum_{k=1}^{\infty} f'_k(x)$, viz. that the differentiated series is convergent, almost everywhere in $[a, b]$. Thus the series

$$\sum_{k=1}^{\infty} \{s'(x) - s'_{n(k)}(x)\}$$

is convergent almost everywhere in (a, b). Therefore

$$s'(x) - s'_{n(k)}(x) \to 0 \quad \text{as } k \to \infty.$$

But the sequence $\{s'_n(x)\}$ is non-decreasing in n. Hence

$$s'(x) - s'_n(x) \to 0 \quad \text{as } n \to \infty.$$

5.9. Exercises

1. Show that the least upper bound of the average metric density of a bounded set E (taken over all sub-intervals of an interval I) is either 0 or 1.

2. If $$E_{mn} = \left\{ x; \frac{3m-2}{3^n} < x < \frac{3m-1}{3^n} \right\},$$

E_n is the union of E_{mn} for $m = 1, 2, ..., 3^{n-1}$, and E is the union of the enumerable collection $E_1, E_2 ...$, show that Cantor's set C is the complement of E with respect to $[0, 1]$.

3. Show that a necessary and sufficient condition that a set of points E should have measure zero is that there should exist an enumerable collection of intervals $I_1, I_2, ...$ of finite total length $\sum_{n=1}^{\infty} |I_n|$ such that each point of E is interior to an infinite number of these intervals (Riesz–Nagy 1953, p. 6).

4. If $f(x)$ is non-increasing and not necessarily continuous in (a, b), show that the limits
$$f(x+0) = \lim_{h \to 0} f(x+h) \quad (h > 0),$$

$$f(x-0) = \lim_{h \to 0} f(x-h) \quad (h > 0)$$

exist at each point x in (a, b).
  Define $f(a-0)$ and $f(b+0)$ to be $f(a)$ and $f(b)$ respectively.
  Let $F(x) = \max\{f(x-0), \ f(x), \ f(x+0)\}$.

A point $x = s$ in (a, b) is said to be dominated if there exists a point $x = \xi$ such that
$$a \leqslant s < \xi \leqslant b$$
and
$$F(s) < f(\xi)$$

Show that the dominated points form an open set, E.

If (a_k, b_k) is an open interval belonging to E, prove that
$$f(a_k + 0) \leqslant F(b_k).$$

Show that the points at which $f(x) \neq F(x)$ are enumerable. Hence deduce that $f(x)$ is differentiable almost everywhere in (a, b).

6 Geometric measure of outer sets and inner sets

6.1. Introduction

It appears from the considerations advanced in § 2.4 that the whole problem of the integration of bounded functions $f(x)$ can be reduced to the problem of the integration of the indicators $\alpha(x, t, f)$ of the sets of points at which $f(x) > t$.

We therefore embark on a systematic method of integrating indicators, guided by the descriptive definition of an integral developed in Chapter 2.

When the integral $\quad \int \alpha(x)\, \mathrm{d}x$

has been defined for an indicator α, its value will be a generalization of the length of an interval that we shall call the *geometric measure* $g(\alpha)$ of the set α.

To define this integral we shall employ the method of bracketing (§ 3.2) and shall therefore need to construct outer and inner bracketing functions $\mu(x)$, $\lambda(x)$, such that $\lambda(x) \leqslant \alpha(x) \leqslant \mu(x)$. These bracketing functions will be the indicators of certain *outer* and *inner* sets of points, which we now proceed to define.

6.2. Elementary sets

The theory of measure in one dimension as given by Lebesgue and by de la Vallée-Poussin is expressed almost entirely in terms of open and closed sets. The exception is the fundamental union–intersection theorem (6.2.5 and 6.3.9), which requires the use of general intervals that may or may not include either of their extremities. Kolmogorov and Fomin (1961) and Williamson

(1962) have simplified the original Lebesgue theory by the systematic use of general intervals throughout the theory, and we shall follow their example in order to give explicit proofs of the basic theorems in a form that can readily be extended to spaces of two or more dimensions.

DEFINITION 6.2.1. In one dimension an 'interval' is a set of points x such that $a \prec x \prec b$, where the symbol $\prec$, introduced by Williamson, represents either $<$ or $\leqslant$, and a, b are finite or infinite numbers. In particular a single point is an interval.

DEFINITION 6.2.2. An 'elementary set', with indicator σ, is the union of a finite number of disjoint intervals, with indicators $\sigma_1, \sigma_2, ..., \sigma_n$

so that $$\sigma_j \sigma_k = 0 \quad \text{if } j \neq k,$$
and $$\sigma = \sigma_1 + \sigma_2 + ... + \sigma_n.$$

THEOREM 6.2.1. *The intersection, union, difference, and symmetric difference of two elementary sets are also elementary sets.*

The intersection of two intervals

$$(a \prec x \prec b) \quad \text{and} \quad (p \prec x \prec q)$$

is an interval of the form

$$\max(a, p) \prec x \prec \min(b, q).$$

Now if σ and τ are elementary sets and

$$\sigma = \sum \sigma_s, \quad \tau = \sum \tau_t,$$

where $\{\sigma_s\}$ and $\{\tau_t\}$ are each disjoint intervals, then the intersection of σ and τ is
$$\sigma\tau = \sum \sigma_s \tau_t.$$

Now $\sigma_s \tau_t$ is an interval and

$$(\sigma_p \tau_q)(\sigma_s \tau_t) = (\sigma_p \sigma_s)(\tau_q \tau_t) = 0$$

unless $$p = s \quad \text{and} \quad q = t.$$

Hence $\sigma\tau$ is the union of a finite number of disjoint intervals and is therefore an elementary set.

If σ is the union of a finite number of disjoint intervals

$$\sigma_s = (a_s \prec x \prec b_s),$$

we can enumerate these so that

$$a_1 \prec b_1 \prec a_2 \prec b_2 \prec ... \prec a_n \prec b_n.$$

Hence if σ is covered by an interval $\omega = (p \prec x \prec q)$, the complement of σ with respect to ω is the finite number of disjoint intervals

$$(p \prec x \prec a_1), \quad (b_1 \prec x \prec a_2), \quad ..., \quad (b_n \prec x \prec q_n)$$

(some of which may be empty sets). Therefore the complement $\omega - \sigma$ is an elementary set. In particular, $1 - \omega$ is an elementary set.

If σ and τ are any two elementary sets, then $1 - \sigma$ and $1 - \tau$ are elementary sets, together with their intersection $(1 - \sigma)(1 - \tau)$ and the union of σ and τ,

$$\sigma \cup \tau = 1 - (1 - \sigma)(1 - \tau).$$

Also the differences

$$\sigma - \sigma\tau \quad \text{and} \quad \tau - \sigma\tau$$

are clearly elementary sets, and the symmetric difference

$$\sigma \bigtriangleup \tau = \sigma \cup \tau - \sigma\tau.$$

Thus elementary sets form an algebra in which 'addition' is defined by union and 'multiplication' by intersection.

THEOREM 6.2.2. *If σ and τ are any pair of elementary sets then there exists a finite collection of disjoint intervals $\{\gamma_s\}$ such that*

$$\sigma = \sum_s a_s \gamma_s,$$

$$\tau = \sum_t b_t \gamma_t,$$

$$\sigma \cup \tau = \sum_s \gamma_s,$$

and each coefficient, a_s or b_t, is either 0 or 1.

Now

$$\sigma = (\sigma - \sigma\tau) + \sigma\tau,$$

$$\tau = (\tau - \sigma\tau) + \sigma\tau,$$

and the elementary sets

$$(\sigma - \sigma\tau), \quad (\tau - \sigma\tau), \quad \text{and} \quad \sigma\tau$$

are disjoint. Let $\{\gamma_s\}$ be the finite collection of disjoint intervals in these elementary sets. Then

$$\sigma = \sum a_s \gamma_s, \quad \tau = \sum_t b_t \gamma_t,$$

and

$$\sigma \cup \tau = (\sigma - \sigma\tau) + (\tau - \sigma\tau) + \sigma\tau$$

$$= \sum_s \gamma_s,$$

where each coefficient is either 0 or 1.

DEFINITION 6.2.3. The geometric measure $g(\alpha)$ of an interval α is its length, and is zero if α is a single point, or the empty set $\emptyset$.

The geometric measure of an elementary set should clearly be defined in terms of the geometric measures of its component intervals, but we must take note of the fact that an elementary set can be resolved into the union of disjoint intervals in an infinite number of ways. For example, the interval $(0 \leqslant x \leqslant 1)$ can be expressed as the union of the intervals $(0 \leqslant x < s)$, $(x = s)$, and $(s < x \leqslant 1)$. We therefore need the following theorems.

THEOREM 6.2.3. *If the interval σ is the union of a finite number of disjoint intervals $\{\sigma_s\}$ then*

$$g(\sigma) = \sum_s g(\sigma_s).$$

For we can enumerate the intervals so that σ_{s+1} lies to the right of σ_s and adjoins σ_s.

THEOREM 6.2.4. *If the elementary set σ has two different representations,*

$$\sigma = \sum_s \alpha_s \quad and \quad \tau = \sum_t \beta_t,$$

as the union of a finite number of disjoint intervals, then

$$\sum_s g(\alpha_s) = \sum_t g(\beta_t).$$

For

$$\sigma = \sigma^2 = \sum_{s,t} \alpha_s \beta_t$$

and

$$\alpha_s = \alpha_s \sigma = \sum_t \alpha_s \beta_t.$$

Hence, by Theorem 6.2.3,

$$g(\alpha_s) = \sum_t g(\alpha_s \beta_t),$$

whence
$$\sum_s g(\alpha_s) = \sum_{s,t} g(\alpha_s \beta_t).$$

Similarly
$$\sum_t g(\beta_t) = \sum_{t,s} g(\alpha_s \beta_t),$$

whence
$$\sum_s g(\alpha_s) = \sum_t g(\beta_t).$$

We can now frame

DEFINITION 6.2.4. The geometric measure $g(\alpha)$ of an elementary set α is
$$g(\alpha) = \sum_s g(\alpha_s),$$
where $\alpha_1, \alpha_2,...$ are the components of α in any finite representation.

The main instrument in proving theorems about geometric measure is the 'union–intersection' theorem, the simplest form of which is the following.

THEOREM 6.2.5. *If α and β are each elementary sets then*
$$g(\alpha \cup \beta) + g(\alpha\beta) = g(\alpha) + g(\beta).$$

By Theorem 6.2.2, α and β can be represented in the form
$$\alpha = \sum_s a_s \gamma_s, \quad \beta = \sum_t b_t \gamma_t$$
where $\{\gamma_s\}$ is a finite collection of disjoint intervals, and each coefficient, a_s or b_t, is either 0 or 1.

Hence
$$g(\alpha) = \sum_s a_s g(\gamma_s), \quad g(\beta) = \sum_t b_t g(\gamma_t),$$
$$g(\alpha\beta) = \sum_s a_s b_s g(\gamma_s),$$
$$g(\alpha \cup \beta) = \sum_s (a_s + b_s - a_s b_s) g(\gamma_s).$$

Whence the theorem follows at once.

COROLLARIES. (i) *If α and β are disjoint elementary sets, then*
$$g(\alpha \cup \beta) = g(\alpha) + g(\beta).$$

(ii) *By induction it follows that if* $\alpha_1, \alpha_2, ..., \alpha_n$ *is a finite collection of disjoint elementary sets, then*

$$g\left(\bigcup_{s=1}^{n} \alpha_s \right) = g(\alpha_1) + g(\alpha_2) + ... + g(\alpha_n).$$

(iii) *If σ and τ are elementary sets and σ is covered by τ, then*

$$g(\sigma) \leqslant g(\tau).$$

For, on writing $\alpha = \sigma$, $\beta = \tau - \sigma$ in the union–intersection theorem, we find that

$$\alpha \cup \beta = \tau, \qquad \alpha\beta = \varnothing,$$

and

$$g(\tau) = g(\sigma) + g(\tau - \sigma) \geqslant g(\sigma).$$

6.3. Bounded outer sets

In the terminology that we have just adopted the theory of measure before Lebesgue may be briefly described as follows. The 'content' of a set of points α was defined as the greatest lower bound of the geometric measures $g(\sigma)$ of all elementary sets σ covering α. But this is clearly a rather crude measure for, according to this definition, both the rational points and the irrational points in the interval $(0, 1)$ have the same content— unity, although the irrational points are vastly more numerous.

In the Lebesgue theory the concept of 'content' is replaced by the concept of 'measure', and this is achieved by covering the set of points α, not by a *finite* collection of disjoint intervals, but by an *enumerable* collection of disjoint intervals. This provides a much finer measure of a set of points, which we now proceed to explore.

DEFINITION 6.3.1. An *outer set* is the union of an enumerable collection of disjoint intervals, called a *representation* of the outer set.

THEOREM 6.3.1. *The union of an enumerable collection of disjoint outer sets is an outer set.*

If the outer set α_n is the union of the enumerable disjoint intervals $\{\sigma_{np}\}$ ($p = 1, 2, ...$) then α, the union of the outer sets $\{\alpha_n\}$ is the union of the intervals $\{\sigma_{np}\}$ ($n, p = 1, 2, ...$).

Now $$\sigma_{np}\sigma_{nq} = 0 \quad \text{if } p \neq q.$$

Also $$\sigma_{mp} \leqslant \alpha_m,$$

whence $$\sigma_{mp} = \alpha_m \sigma_{mp} \quad \text{(by Theorem 4.1.2),}$$

and
$$\sigma_{mp}\sigma_{nq} = (\alpha_m \sigma_{mp})(\alpha_n \sigma_{nq})$$
$$= (\alpha_m \alpha_n)(\sigma_{mp}\sigma_{nq})$$
$$= 0 \quad \text{if } m \neq n.$$

Hence the collection of intervals (σ_{np}) is disjoint, whence α is an outer set.

THEOREM 6.3.2. *The union of an enumerable collection of (not necessarily disjoint) intervals is an outer set.*

If the set α is the union of the bounded, enumerable, non-disjoint intervals $\{\alpha_n\}$ $(n = 1, 2,...)$ then by the covering Theorem 4.4.1, α is also the union of the enumerable, disjoint sets $\{\beta_n\}$ where
$$\beta_1 = \alpha_1$$

and $$\beta_n = \alpha_n(1-\alpha_1)(1-\alpha_2)...(1-\alpha_{n-1}) \quad (q > 1).$$

But, by Theorem 6.2.1, $(1-\alpha_1), (1-\alpha_2),..., (1-\alpha_{n-1})$ are elementary sets, whence β_n is an elementary set, i.e. the union of a finite number of disjoint intervals. Hence α is the union of an enumerable collection of disjoint intervals, and is therefore an outer set.

THEOREM 6.3.3. *The intersection of any two outer sets is an outer set.*

If the outer sets α and β are respectively the unions of the enumerable collections of disjoint intervals $\{\alpha_m\}$ and $\{\beta_n\}$, then the intersection $\alpha\beta$ is the union of the enumerable intervals $\{\alpha_m\beta_n\}$. But
$$(\alpha_m\beta_n)(\alpha_p\beta_q) = (\alpha_m\alpha_p)(\beta_n\beta_q)$$
$$= 0 \quad \text{unless } m = p \quad \text{and} \quad n = q.$$

Hence the intervals $\alpha_m\beta_n$ are disjoint, and therefore the intersection $\alpha\beta$ is an outer set.

THEOREM 6.3.4. *The union of any two outer sets is an outer set.*

In the notation of Theorem 6.3.3, the union $\alpha \cup \beta$ of the outer sets α and β is the union of the enumerable collection of intervals $\{\alpha_m\}$ and $\{\beta_n\}$. Hence, by Theorem 6.3.2, $\alpha \cup \beta$ is an outer set.

We must next establish the existence of geometric measures for outer sets.

THEOREM 6.3.5. *If σ is a bounded outer set with enumerable disjoint components $\{\sigma_n\}$ then the series*

$$\sum_{1}^{\infty} g(\sigma_n)$$

is convergent.

If σ is covered by an interval ω, then the union τ_n of the finite collection of disjoint intervals $\sigma_1, \sigma_2, \dots, \sigma_n$ is an elementary set also covered by ω. Hence, by Corollary (iii) to Theorem 6.2.5,

$$\sum_{s=1}^{n} g(\sigma_s) = g(\tau_n) \leqslant g(\omega).$$

Hence the series $\sum_{1}^{\infty} g(\sigma_n)$ is bounded. Since each term is non-negative, the series is therefore convergent.

THEOREM 6.3.6. *If σ and τ are bounded outer sets with the representations*

$$\sigma = \sum_{s=1}^{\infty} \alpha_s \quad and \quad \tau = \sum_{t=1}^{\infty} \beta_t$$

and if σ is covered by τ, then

$$\sum_{s=1}^{\infty} g(\alpha_s) \leqslant \sum_{t=1}^{\infty} g(\beta_t).$$

We shall compare the elementary set

$$\sigma_m = \sum_{s=1}^{m} \alpha_s$$

and the outer set
$$\sum_{t=1}^{\infty} \beta_t.$$

We replace each interval α_s by a *closed* interval α_s^* covered by α_s and only slightly smaller so that

$$\alpha_s^* \leqslant \alpha_s,$$

and $\qquad g(\alpha_s) - \epsilon/2^s \leqslant g(\alpha_s^*) \leqslant g(\alpha_s),$

for $s = 1, 2, ..., m, \epsilon$ being an arbitrary positive tolerance (if α_s is a point, then α_s^* is an empty set!). The union of the closed, disjoint intervals $\{\alpha_s^*\}$ forms a closed set.

We replace each interval β_t by an open interval β_t^* covering β_t and only slightly larger so that

$$\beta_t \leqslant \beta_t^*$$

and $\qquad g(\beta_t) \leqslant g(\beta_t^*) \leqslant g(\beta_t) + \epsilon/2^t.$

The closed set $\sum_s \alpha_s^*$ is covered by the enumerable open intervals $\{\beta_t^*\}$. Hence by the Heine–Borel covering theorem (4.4.2) the closed set $\sum_s \alpha_s^*$ is also covered by the union of a finite number of open intervals $\beta_{t_1}^*, \beta_{t_2}^*, ..., \beta_{t_n}^*$.

Therefore, by Corollary (iii) to Theorem 6.2.5,

$$\sum_{s=1}^{m} g(\alpha_s^*) \leqslant \sum_{m=1}^{n} g(\beta_{t_n}^*).$$

Hence

$$\sum_{s=1}^{m} g(\alpha_s) \leqslant \sum_{s=1}^{m} g(\alpha_s^*) + \sum_{s=1}^{m} \epsilon/2^s \leqslant \sum_{m=1}^{n} g(\beta_{t_m}^*) + \epsilon \leqslant \sum_{t=1}^{\infty} g(\beta_t) + 2\epsilon.$$

Since this is true for all m and ϵ,

$$\sum_{s=1}^{\infty} g(\alpha_s) \leqslant \sum_{t=1}^{\infty} g(\beta_t).$$

COROLLARY. *If σ is a bounded outer set with two representations*

$$\sigma = \sum_{s=1}^{\infty} \alpha_s \quad and \quad \sigma = \sum_{t=1}^{\infty} \beta_t,$$

then $\qquad \displaystyle\sum_{s=1}^{\infty} g(\alpha_s) = \sum_{t=1}^{\infty} g(\beta_t).$

For σ is covered by σ, whence, by the theorem,

$$\sum_{s=1}^{\infty} g(\alpha_s) \leqslant \sum_{t=1}^{\infty} g(\beta_t).$$

Similarly
$$\sum_{t=1}^{\infty} g(\beta_t) \leqslant \sum_{s=1}^{\infty} g(\alpha_s),$$

and therefore
$$\sum_{s=1}^{\infty} g(\alpha_s) = \sum_{t=1}^{\infty} g(\beta_t).$$

We can now define the geometric measure of any bounded outer set α.

DEFINITION 6.3.2. *If* $\sum_{s=1}^{\infty} \alpha_s$ *is any representation of a bounded, outer set* α, *then the geometric measure of* α *is*

$$g(\alpha) = \sum_{s=1}^{\infty} g(\alpha_s).$$

THEOREM 6.3.7. *If* σ *and* τ *are bounded outer sets and* σ *is covered by* τ, *then*
$$g(\sigma) \leqslant g(\tau).$$

This follows at once from Definition 6.3.2 and Theorem 6.3.6.

We can now generalize the union–intersection theorem for an enumerable collection of bounded outer sets. First we consider disjoint sets.

THEOREM 6.3.8. *If* α_1, α_2,... *are enumerable, disjoint outer sets with a bounded union* α, *then* α *is an outer set and*
$$g(\alpha) = g(\alpha_1) + g(\alpha_2) + \dots .$$

For if α_s is the union of enumerable disjoint intervals $\{\alpha_{st}\}$, then $\alpha_s \leqslant \alpha$, whence α_s is bounded and, by Definition 6.3.2,

$$g(\alpha_s) = \sum_{t=1}^{\infty} g(\alpha_{st}).$$

But α is the union of the disjoint intervals $\{\alpha_{st}\}$. Therefore, by the theorem on the derangement of double series with non-negative terms,

$$g(\alpha) = \sum_{s,t=1}^{\infty} g(\alpha_{st}) = \sum_{s=1}^{\infty} g(\alpha_s).$$

COROLLARY. *If* β_1, β_2,... *are uniformly bounded, enumerable elementary sets such that*
$$\beta_n \leqslant \beta_{n+1} \quad (n = 1, 2, \dots),$$
then
$$g(\beta_n) \to g(\beta) \quad as \quad n \to \infty$$
where
$$\beta = \lim_{n \to \infty} \beta_n.$$

For
$$\beta = \beta_1 + \alpha_1 + \alpha_2 + \ldots,$$
where
$$\alpha_n = \beta_{n+1} - \beta_n,$$
and
$$\beta_1, \alpha_1, \alpha_2, \ldots$$
are disjoint elementary sets. to which we can apply the main theorem.

Secondly we consider outer sets which are not necessarily disjoint.

THEOREM 6.3.9 (the union–intersection theorem for outer sets). *If α and β are each bounded outer sets, then*
$$g(\alpha \cup \beta) + g(\alpha\beta) = g(\alpha) + g(\beta).$$

By Theorems 6.3.3 and 6.3.4, $\alpha\beta$ and $\alpha \cup \beta$ are each outer sets.

Let $\alpha = \sum_s \alpha_s$ and $\beta = \sum_t \beta_t$ be representations of α and β in terms of enumerable, disjoint intervals.

Let
$$\sigma_m = \sum_{s=1}^m \alpha_s \quad \text{and} \quad \tau_n = \sum_{t=1}^n \beta_t.$$
Then, by Definition 6.3.2,
$$g(\sigma_m) = \sum_{s=1}^m g(\alpha_s) \quad \text{and} \quad g(\tau_n) = \sum_{t=1}^n g(\beta_t)$$
and
$$g(\sigma_m \tau_n) = \sum_{s,t=1}^{m,n} g(\alpha_s \beta_t).$$
Hence, as $m, n \to \infty$,
$$g(\sigma_m) \to g(\alpha), \quad g(\tau_n) \to g(\beta), \quad g(\sigma_m \tau_n) \to g(\alpha\beta).$$
Also, if
$$\gamma_n = \sigma_n \cup \tau_n,$$
then
$$\gamma_n \leqslant \gamma_{n+1}$$
and
$$\gamma_n \to \alpha \cup \beta \quad \text{as } n \to \infty.$$
Therefore, by the corollary to Theorem 6.3.8,
$$g(\gamma_n) \to g(\alpha \cup \beta).$$

But, by the union–intersection theorem for elementary sets (6.2.5)
$$g(\sigma_n \cup \tau_n) + g(\sigma_n \tau_n) = g(\sigma_n) + g(\tau_n).$$
Now let $n \to \infty$. Then
$$g(\alpha \cup \beta) + g(\alpha\beta) = g(\alpha) + g(\beta).$$

As in the case of Theorem 6.2.5 there are a number of obvious corollaries to this theorem.

COROLLARY. (i) *If α and β are bounded disjoint outer sets, then*
$$g(\alpha \cup \beta) = g(\alpha) + g(\beta).$$

(ii) *By induction it follows that if $\alpha_1, \alpha_2, ..., \alpha_n$ is a finite collection of bounded disjoint outer sets, then*
$$g\left(\bigcup_{s=1}^{n} \alpha_s \right) = g(\alpha_1) + g(\alpha_2) + ... + g(\alpha_n).$$

(iii) *If $\{\alpha_n\}$ is any finite collection of bounded outer sets with union α then $g(\alpha_1) + g(\alpha_2) \geqslant g(\alpha_1 \cup \alpha_2)$ and by induction*
$$\sum_{s=1}^{n} g(\alpha_s) \geqslant g(\alpha).$$

6.4. Unbounded outer sets

The definition of the geometric measure of an unbounded outer set σ can be achieved only by considering the family of sets $\{\sigma_s\}$, where σ_s is the intersection of σ and the open interval
$$-s < x < s,$$
s being any positive number. Thus, as $s \to \infty$ these open intervals steadily expand to infinity.

THEOREM 6.4.1. *As $s \to \infty$, $g(\sigma_s)$ tends to a unique limit λ.*

For, if $s < t$, then $\qquad \sigma_s \leqslant \sigma_t$,
and, by Theorem 6.3.7,
$$g(\sigma_s) \leqslant g(\sigma_t).$$
Hence the function $g(\sigma_s)$ converges to a unique limit λ as $s \to \infty$, which may be finite or infinite.

DEFINITION 6.4.1. The geometric measure $g(\sigma)$ of the unbounded outer set σ is the limit,
$$g(\sigma) = \lim_{s \to \infty} g(\sigma_s).$$

6.5. The principle of complementarity

So far we have defined and studied only outer sets of points—the indicators of which are to be the upper bracketing functions in our definition of the Lebesgue measure. We must now examine inner sets—the indicators of which are to be the lower bracketing functions. In the original Lebesgue theory the outer sets were open sets and the inner sets were closed sets. We need to modify this theory in view of the generalization of outer sets that we adopted in § 6.3.

The motivation of the Lebesgue theory was the 'principle of complementarity', which we can describe as follows.

If σ and τ are complementary with respect to an interval ω then

$$\sigma + \tau = \omega \quad \text{and} \quad \sigma\tau = 0.$$

Now let μ be an outer set covering σ and ν be an outer set covering τ. Then

$$\sigma \leqslant \mu, \quad \tau \leqslant \nu$$

and

$$\omega - \nu \leqslant \omega - \tau = \sigma.$$

Hence σ is bracketed by the sets

$$\lambda \equiv \omega - \nu \quad \text{and} \quad \mu.$$

The lower bracketing set λ is the complement of the upper bracketing set ν with respect to the interval ω. We shall take the upper bracketing sets to be 'outer sets' covering the prescribed set σ; and we shall take the lower bracketing sets to be 'inner sets' which are the complements of outer sets with respect to an interval.

But it is necessary to give a self-consistent definition of inner sets λ and of their geometric measure $g(\lambda)$ and to prove that if a given set σ is bracketed by an outer set μ and an inner set λ then $g(\lambda) \leqslant g(\mu)$. We therefore proceed to establish the necessary definitions and theorems.

6.6. Inner sets

THEOREM 6.6.1. *If λ is the complement of an outer set μ_1 with respect to an interval ω_1, then it is also the complement of some outer set μ_2 with respect to any other interval ω_2 which covers λ.*

(1) We first prove that λ is the complement of an outer set μ with respect to the interval $\omega = \omega_1 \omega_2$.

Since λ is covered by ω_1 and by ω_2, it is also covered by ω. Let

$$\omega = \lambda + \mu.$$

Then
$$\mu = \omega - \lambda = \omega - (\omega_1 - \mu_1)$$
$$= \mu_1 - (\omega_1 - \omega).$$

Now $\omega = \omega_1 \omega_2 \leqslant \omega_1$. Thus $\omega_1 - \omega$ is a set, which clearly consists of one or two intervals.

Also
$$\omega_1 - \omega \leqslant \mu_1,$$

whence
$$(\omega_1 - \omega)\mu_1 = (\omega_1 - \omega),$$

and
$$\mu = \mu_1(1 - \omega_1 + \omega).$$

But $1 - \omega_1 + \omega$ is a set consisting of two or three intervals, and μ_1 is the union of an enumerable collection of disjoint intervals. Hence so also is μ, i.e. μ is an outer set.

(2) In the general case, let $\omega_2 = \lambda + \mu_2$, whence

$$\mu_2 = \omega_2 - \omega + \mu.$$

But
$$\mu \leqslant \omega \leqslant \omega_2,$$

and
$$\mu = \mu\omega = \mu\omega_2,$$

so that
$$(\omega_2 - \omega)\mu = 0.$$

Thus μ_2 is the union of two disjoint sets, viz.:

(i) $\omega_2 - \omega$, which consists of one or two intervals, and
(ii) μ, which we have proved to be an outer set.

Therefore μ_2 is also an outer set.

DEFINITION 6.6.1. Any set λ that is the complement of an outer set μ with respect to some interval ω is an 'inner set'.

DEFINITION 6.6.2. The geometric measure of an inner set λ with respect to a finite interval ω covering λ is defined to be

$$g_\omega(\lambda) = g(\omega) - g(\omega - \lambda).$$

THEOREM 6.6.2. *If $g_{\omega_1}(\lambda)$ and $g_{\omega_2}(\lambda)$ are the geometric measures of an inner set λ with respect to two intervals ω_1 and ω_2 each covering λ, then*

$$g_{\omega_1}(\lambda) = g_{\omega_2}(\lambda).$$

We apply the union–intersection theorem (6.3.9) to the sets

$$\alpha = \omega_1 \quad \text{and} \quad \beta = \omega_2 - \lambda.$$

Their union is

$$\alpha \cup \beta = \omega_1 + (\omega_2 - \lambda) - (\omega_1 \omega_2 - \omega_1 \lambda).$$

But λ is covered by ω_1, whence

$$\omega_1 \lambda = \lambda,$$

and $\qquad \alpha \cup \beta = \omega_1 + \omega_2 - \omega_1 \omega_2 = \omega_1 \cup \omega_2.$

The intersection of α and β is

$$\alpha\beta = \omega_1(\omega_2 - \lambda)$$
$$= \omega_1 \omega_2 - \lambda.$$

Hence $\qquad g(\omega_1) + g(\omega_2 - \lambda) = g(\alpha) + g(\beta)$
$$= g(\alpha \cup \beta) + g(\alpha\beta)$$
$$= g(\omega_1 \cup \omega_2) + g(\omega_1 \omega_2 - \lambda).$$

Similarly $\qquad g(\omega_2) + g(\omega_1 - \lambda) = g(\omega_1 \cup \omega_2) + g(\omega_1 \omega_2 - \lambda).$

Therefore $\qquad g(\omega_1) + g(\omega_2 - \lambda) = g(\omega_2) + g(\omega_1 - \lambda)$

and $\qquad g(\omega_1) - g(\omega_1 - \lambda) = g(\omega_2) - g(\omega_2 - \lambda).$

DEFINITION 6.6.3. The geometric measure of a bounded inner set λ is $g(\lambda) = g(\omega) - g(\omega - \lambda)$, where ω is any finite interval covering λ.

THEOREM 6.6.3. *If the bounded set σ is bracketed by an outer set μ and an inner set λ then $g(\lambda) \leqslant g(\mu)$.*

For $\qquad\qquad\qquad \lambda \leqslant \sigma \leqslant \mu$

and there is a finite interval ω covering λ and μ such that

$$\nu = \omega - \lambda$$

is an outer set (by Definition 6.6.1). Hence

$$\mu \cup \nu = \mu + \nu - \mu\nu$$
$$= \mu + \omega - \lambda - \mu\omega + \mu\lambda$$
$$= \omega$$

since $\qquad\qquad \mu = \mu\omega \quad \text{and} \quad \lambda = \mu\lambda.$

Therefore by the union–intersection theorem (6.3.9)

$$g(\omega) = g(\mu \cup \nu) = g(\mu) + g(\nu) - g(\mu\nu) \leqslant g(\mu) + g(\nu).$$

But
$$g(\lambda) = g(\omega) - g(\nu),$$
whence
$$g(\lambda) \leqslant g(\mu).$$

It is almost trivial, but nevertheless necessary, to establish the following theorem.

THEOREM 6.6.4. *If σ is any bounded set, then there always exist outer sets μ and inner sets λ such that $\lambda \leqslant \sigma \leqslant \mu$.*

Since σ is bounded there is an interval μ covering σ, and any point of σ is an interval λ covered by σ. λ and μ are inner and outer sets bracketing σ.

6.7. Exercises

1. Show that a single point $x = a$ is both an outer set and an inner set. Show that the empty set is both an outer set and an inner set.

2. A function $f(x)$ is semi-continuous at $x = \xi$

(i) on the right, if $f(\xi+h) \to f(\xi)$ as $h \to 0$ for $h > 0$,   or
(ii) on the left, if $f(\xi-h) \to f(\xi)$ as $h \to 0$ for $h < 0$.

Show that $\alpha(x)$ is the indicator of an outer set if $\alpha(x)$ is semi-continuous (on either right or left or both) at each point ξ where $\alpha(\xi) = 1$.

Show that $\alpha(x)$ is the indicator of an inner set if $\alpha(x)$ is semi-continuous at each point ξ where $\alpha(\xi) = 0$.

3. Show that the intersection of a finite or an enumerable collection of inner sets is also an inner set (which may be empty) (see Theorem 6.3.2).

4. Show that the intersection of a finite number of outer sets is also an outer set.

7 Lebesgue measure

7.1. Introduction

In the preceding chapter we have developed the theory of the geometric measure of outer and inner sets and have shown that, if σ is any bounded set of points, then there exist outer and inner sets μ and λ such that

$$\lambda \leqslant \sigma \leqslant \mu$$

and $$g(\lambda) \leqslant g(\mu).$$

Our guiding principles have been those originally adopted by Lebesgue, viz.:

(i) the use of *enumerable* collections of disjoint intervals $\{\mu_n\}$ for the upper bracketing function μ, and

(ii) the use of the principle of *complementarity* to define the lower bracketing function λ.

Lebesgue himself never gave a complete and formal account of his theory of measure and integration and the first systematic treatment was given by de la Vallée-Poussin. In his account which we follow here, the central position is occupied by the 'union–intersection theorem'.

In this chapter we develop the theory of measure as the theory of the integration of the indicators $\alpha(x)$ of bounded linear sets of points.

7.2. Outer and inner measure

DEFINITION 7.2.1. The outer measure $m^*(\sigma)$ of a set of points σ is the greatest lower bound of the geometric measures $g(\mu)$ of outer sets μ covering σ, i.e.

$$m^*(\sigma) = \inf g(\mu) \quad \text{for } \mu \geqslant \sigma.$$

DEFINITION 7.2.2. The inner measure $m_*(\sigma)$ of a set of points σ is the least upper bound of the geometric measures $g(\lambda)$ of the inner sets λ covered by σ, i.e.

$$m_*(\sigma) = \sup g(\lambda) \quad \text{for } \lambda \leqslant \sigma.$$

THEOREM 7.2.1. *The inner measure of any set σ is not greater than its outer measure, i.e.*

$$m_*(\sigma) \leqslant m^*(\sigma).$$

For, with the notation of the preceding definitions,

$$g(\lambda) \leqslant g(\mu) \quad \text{(by Theorem 6.6.3)},$$

whence $\qquad m_*(\sigma) = \sup g(\lambda) \leqslant \inf g(\mu) = m^*(\sigma).$

The union–intersection theorem (6.3.9) for the geometric measure of outer sets can now be generalized to the outer and inner measures of any pair of sets.

THEOREM 7.2.2. *If σ and τ are any pair of sets, then*

$$m^*(\sigma) + m^*(\tau) \geqslant m^*(\sigma \cup \tau) + m^*(\sigma\tau).$$

From Definition 7.2.1, if ϵ is any prescribed positive tolerance, there exist outer sets α and β such that

$$\sigma \leqslant \alpha, \qquad \tau \leqslant \beta$$

and $\qquad m^*(\sigma) \geqslant g(\alpha) - \epsilon, \qquad m^*(\tau) \geqslant g(\beta) - \epsilon.$

Now
$$\sigma \cup \tau = 1 - (1-\sigma)(1-\tau)$$
$$\leqslant 1 - (1-\alpha)(1-\beta)$$
$$= \alpha \cup \beta,$$

and $\qquad \sigma\tau \leqslant \alpha\beta.$

But $\alpha \cup \beta$ and $\alpha\beta$ are outer sets. Therefore

$$m^*(\sigma \cup \tau) \leqslant g(\alpha \cup \beta)$$

and $\qquad m^*(\sigma\tau) \leqslant g(\alpha\beta).$

By the union–intersection theorem (6.3.9)

$$g(\alpha \cup \beta) + g(\alpha\beta) = g(\alpha) + g(\beta),$$

whence

$$m^*(\sigma \cup \tau) + m^*(\sigma\tau) \leqslant g(\alpha) + g(\beta) \leqslant m^*(\sigma) + m^*(\tau) + 2\epsilon.$$

Since this is true for all $\epsilon > 0$ it follows that

$$m^*(\sigma \cup \tau) + m^*(\sigma\tau) \leqslant m^*(\sigma) + m^*(\tau).$$

THEOREM 7.2.3. *If σ and τ are any pair of sets, then*

$$m_*(\sigma \cup \tau) + m_*(\sigma\tau) \geqslant m_*(\sigma) + m_*(\tau).$$

Let ω be any finite interval covering σ and τ, and

$$\sigma' = \omega - \sigma, \quad \tau' = \omega - \tau.$$

Then
$$\sigma' \cup \tau' = \omega - \sigma\tau$$

and
$$\sigma'\tau' = \omega - \sigma \cup \tau.$$

Hence
$$m_*(\sigma \cup \tau) = g(\omega) - m^*(\sigma'\tau'),$$

$$m_*(\sigma\tau) = g(\omega) - m^*(\sigma' \cup \tau'),$$

and $\quad m_*(\sigma \cup \tau) + m_*(\sigma\tau) \geqslant 2g(\omega) - m^*(\sigma') - m^*(\tau')$

$$\text{(by Theorem 7.2.2)}$$

$$= m_*(\sigma) + m_*(\tau).$$

COROLLARY. *If $\sigma_1, \sigma_2,..., \sigma_n$ are disjoint sets, then*

$$m_*(\sigma_1 + \sigma_2 + ... + \sigma_n) \geqslant m_*(\sigma_1) + m_*(\sigma_2) + ... + m_*(\sigma_n).$$

For
$$m_*(\sigma_1 + \sigma_2) \geqslant m_*(\sigma_1) + m_*(\sigma_2),$$

and the corollary follows by induction.

THEOREM 7.2.4. *If the set σ is covered by the set τ, i.e. if*

$$\sigma \leqslant \tau,$$

then $\quad m^*(\sigma) \leqslant m^*(\tau) \quad and \quad m_*(\sigma) \leqslant m_*(\tau).$

If ϵ is any prescribed tolerance there exists an outer set ν such that
$$\tau \leqslant \nu \quad and \quad g(\nu) - \epsilon \leqslant m^*(\tau).$$

But
$$\sigma \leqslant \tau \leqslant \nu,$$

whence
$$m^*(\sigma) \leqslant g(\nu) \leqslant m^*(\tau) + \epsilon.$$

Since this is true for each $\epsilon > 0$ it follows that

$$m^*(\sigma) \leqslant m^*(\tau).$$

Also
$$m_*(\sigma) = g(\omega) - m^*(\omega - \sigma),$$
$$m_*(\tau) = g(\omega) - m^*(\omega - \tau),$$

and
$$\omega - \tau \leqslant \omega - \sigma,$$

whence $m_*(\tau) - m_*(\sigma) = m^*(\omega - \sigma) - m^*(\omega - \tau) \geqslant 0,$

by the first part of this theorem.

The following theorem is needed later (§ 7.9, Exercise 4) in constructing a criterion for measurability.

THEOREM 7.2.5. *If σ and τ are any two sets and δ is their symmetric difference,*
$$\delta = \sigma \cup \tau - \sigma\tau$$
$$= \sigma + \tau - 2\sigma\tau,$$

then
$$|m^*(\sigma) - m^*(\tau)| \leqslant m^*(\delta).$$

We note that
$$\tau \cup \delta = \tau + (\sigma + \tau - 2\sigma\tau) - \tau(\sigma + \tau - 2\sigma\tau)$$
$$= \sigma + \tau(1 - \sigma) \geqslant \sigma.$$

Hence, by Theorem 7.2.4,
$$m^*(\sigma) \leqslant m^*(\tau \cup \delta),$$

and by the union–intersection Theorem (7.2.2),
$$m^*(\tau \cup \delta) + m^*(\tau\delta) \leqslant m^*(\tau) + m^*(\delta).$$

Therefore
$$m^*(\sigma) - m^*(\tau) \leqslant m^*(\delta) - m^*(\tau\delta) \leqslant m^*(\delta).$$

Similarly
$$m^*(\tau) - m^*(\sigma) \leqslant m^*(\delta).$$

Whence the theorem follows.

7.3. Lebesgue measure

If λ and μ are inner and outer sets bracketing a given bounded set σ, then
$$g(\mu) - g(\lambda) \geqslant m^*(\sigma) - m_*(\sigma).$$

Thus if $g(\lambda)$ and $g(\mu)$ are to be regarded as lower and upper approximations to the integral of $\sigma(x)$ the tolerance of this bracketing process, $g(\mu) - g(\lambda)$, cannot be less than $m^*(\sigma) - m_*(\sigma)$ and it will not succeed in providing a definition of $\int \sigma(x)\, dx$

unless $m^*(\sigma) = m_*(\sigma)$. Hence we are forced to the following definition.

DEFINITION 7.3.1. A bounded set σ is measurable, in the sense of Lebesgue, if $m^*(\sigma) = m_*(\sigma)$ and the common value of its outer and inner measures is the Lebesgue measure $m(\sigma)$ of σ, i.e.

$$m(\sigma) = m^*(\sigma) = m_*(\sigma).$$

We note at once

THEOREM 7.3.1. *If the set σ bounded by an interval ω is measurable then so also is the complementary set $\tau = \omega - \sigma$, and* $m(\sigma) + m(\tau) = g(\omega)$.

For, by Definition 7.2.2,

$$m_*(\tau) = g(\omega) - m^*(\sigma)$$

and

$$m_*(\sigma) = g(\omega) - m^*(\tau),$$

whence

$$m^*(\tau) - m_*(\tau) = m^*(\sigma) - m_*(\sigma) = 0,$$

and

$$m(\sigma) + m(\tau) = g(\omega).$$

We can now complete the series of union–intersection theorems as follows.

THEOREM 7.3.2. *If σ and τ are bounded measurable sets, then so is their intersection $\sigma\tau$ and their union $\sigma \cup \tau$, and*

$$m(\sigma \cup \tau) + m(\sigma\tau) = m(\sigma) + m(\tau).$$

By the union–intersection theorems for outer and inner measure, (7.2.2) and (7.2.3),

$$m^*(\sigma \cup \tau) + m^*(\sigma\tau) \leqslant m^*(\sigma) + m^*(\tau)$$
$$= m_*(\sigma) + m_*(\tau)$$
$$\leqslant m_*(\sigma \cup \tau) + m_*(\sigma\tau).$$

Hence

$$\{m^*(\sigma \cup \tau) - m_*(\sigma \cup \tau)\} + \{m^*(\sigma\tau) - m_*(\sigma\tau)\} \leqslant 0.$$

But, by Theorem 7.2.1, the bracketed expressions are each non-negative. They are therefore each equal to zero, i.e.

$$m^*(\sigma \cup \tau) = m_*(\sigma \cup \tau)$$

and

$$m^*(\sigma\tau) = m_*(\sigma\tau).$$

Thus the union, $\sigma \cup \tau$, and the intersection, $\sigma\tau$, are each measurable.

Therefore, using the union–intersection theorems again,

$$m(\sigma \cup \tau) + m(\sigma\tau) \leqslant m(\sigma) + m(\tau)$$
$$\leqslant m(\sigma \cup \tau) + m(\sigma\tau),$$

whence $\quad m(\sigma \cup \tau) + m(\sigma\tau) = m(\sigma) + m(\tau).$

COROLLARY. (i) *If σ and τ are disjoint measurable sets then*

$$m(\sigma + \tau) = m(\sigma) + m(\tau).$$

By induction, if $\sigma_1, \sigma_2, ..., \sigma_n$ is a finite collection of disjoint measurable sets then

$$m(\sigma_1 + \sigma_2 + ... + \sigma_n) = m(\sigma_1) + m(\sigma_2) + ... + m(\sigma_n).$$

It is not quite trivial to notice that

(ii) *If σ and τ are measurable and σ covers τ, then $m(\sigma) \geqslant m(\tau)$.*

For $\eta = \sigma(1-\tau)$ is measurable and $m(\sigma) = m(\tau) + m(\eta) \geqslant m(\tau)$ (cf. Theorem 7.2.4).

THEOREM 7.3.3. *If $\{\sigma_n\}$ is any enumerable collection of sets of points with bounded union σ, then*

$$m^*(\sigma) \leqslant \sum_{n=1}^{\infty} m^*(\sigma_n).$$

By Definition 7.2.1, for any prescribed tolerance $\epsilon > 0$, there is an outer set α_n such that

$$\sigma_n \leqslant \alpha_n,$$

and $\quad g(\alpha_n) - \epsilon/2^n \leqslant m^*(\sigma_n).$

The outer set α_n is the union of an enumerable collection of disjoint intervals $\{\alpha_{np}\}$ ($p = 1, 2, ...$), whence α, the union of all the outer sets $\{\alpha_n\}$, is the union of an enumerable collection of intervals $\{\alpha_{np}\}$ ($n, p = 1, 2, ...$). These intervals are not necessarily disjoint, but by the corollary to the covering Theorem 4.4.1 there exists an enumerable collection of disjoint elementary sets $\{\beta_{np}\}$, such that

$$\beta_{np} \leqslant \alpha_{np}$$

and $\quad \bigcup \beta_{np} = \bigcup \alpha_{np}.$

Since $\sigma = \bigcup \sigma_n$ is covered by $\bigcup \alpha_n$, it is also covered by $\bigcup \alpha_{np}$ and by $\bigcup \beta_{np}$ $(n, p = 1, 2, ...)$. Hence

$$m^*(\sigma) \leqslant g\left(\bigcup_{n,p=1}^{\infty} \beta_{np} \right) = \sum_{n,p=1}^{\infty} g(\beta_{np})$$

$$\leqslant \sum_{n,p=1}^{\infty} g(\alpha_{np}) \quad \text{(by Theorem 6.3.7)}$$

$$= \sum_{n=1}^{\infty} g(\alpha_n) \leqslant \sum_{n=1}^{\infty} m^*(\sigma_n) + \sum_{n=1}^{\infty} \epsilon/2^n = \sum_{n=1}^{\infty} m^*(\sigma_n) + \epsilon.$$

This is true for all $\epsilon > 0$, so that the theorem is established.

THEOREM 7.3.4. *If $\{\sigma_n\}$ is any enumerable collection of disjoint sets of points with union σ, then*

$$m_*(\sigma) \geqslant \sum_{n=1}^{\infty} m_*(\sigma_n).$$

By Theorem 7.2.4, since σ covers the union of $\sigma_1, \sigma_2,..., \sigma_n$

$$m_*(\sigma) \geqslant m_*\left(\bigcup_1^n \sigma_p \right)$$

$$\geqslant m_*(\sigma_1) + m_*(\sigma_2) + ... + m_*(\sigma_n),$$

by the corollary Theorem 7.2.3.

Since this is true for all n, it follows that

$$m_*(\sigma) \geqslant \sum_{n=1}^{\infty} m_*(\sigma_n).$$

Finally, by combining Theorems 7.3.3 and 7.3.4 we obtain

THEOREM 7.3.5. *If $\{\sigma_n\}$ is any enumerable collection of disjoint measurable sets, with union σ, then σ is measurable and*

$$m(\sigma) = \sum_{n=1}^{\infty} m(\sigma_n).$$

The corollary to Theorem 7.3.2 states that the Lebesgue measure is an additive functional in the sense that, if $\sigma_1, \sigma_2,..., \sigma_n$ is a finite collection of disjoint measurable sets, then

$$m(\sigma_1 + \sigma_2 + ... + \sigma_n) = m(\sigma_1) + m(\sigma_2) + ... + m(\sigma_n).$$

We have now proved that the Lebesgue measure is *completely* additive in the sense that, if $\{\sigma_n\}$ is an enumerable collection of disjoint sets then their union is measurable and

$$m(\sigma_1 + \sigma_2 + ...) = m(\sigma_1) + m(\sigma_2) + ...$$

The Lebesgue measure $m(\sigma)$ of a measurable set σ is therefore a positive, additive, continuous functional of σ and we shall prove (Theorem 7.4.2) that it satisfies the normalizing condition that if σ is an interval, then $m(\sigma)$ is the length of the interval. Hence the Lebesgue measure $m(\sigma)$ can rightly be taken to be the integral of the indicator σ.

7.4. Examples of measurable sets

The simplest sets of points are finite or enumerable collections of points and we can prove at once

THEOREM 7.4.1. *Any enumerable collection of points E is measurable and has Lebesgue measure zero.*

When E contains only one point ξ it is covered by the interval

$$\xi - \tfrac{1}{2}\epsilon < x < \xi + \tfrac{1}{2}\epsilon$$

with geometric measure ϵ. Hence

$$0 \leqslant m_*(E) \leqslant m^*(E) < \epsilon \quad \text{for all } \epsilon > 0,$$

and

$$m^*(E) - m_*(E) = 0.$$

Thus E is measurable and $m(E) = 0$.

Hence, by the completely additive property of Lebesgue measure (Theorem 7.3.5), an enumerable collection E of points $\{E_n\}$ is measurable and

$$m(E) = \sum_{n=1}^{\infty} m(E_n) = 0.$$

Thus a set of points with 'zero one-dimensional' measure, according to Definition 5.2.1, has Lebesgue measure zero in accordance with Definition 7.3.1.

In order of increasing complexity the next set of points to be considered is the interval.

THEOREM 7.4.2. *Any interval α is measurable and its Lebesgue measure is equal to its geometric measure, i.e.*

$$m(\alpha) = g(\alpha).$$

Since α is covered by α it follows from Theorem 7.2.1 that

$$m^*(\alpha) \leqslant g(\alpha).$$

Also, if α is covered by an interval ω then

$$m_*(\alpha) \geqslant g(\omega) - g(\omega - \alpha).$$

Hence

$$m^*(\alpha) - m_*(\alpha) \leqslant g(\alpha) + g(\omega - \alpha) - g(\omega) = 0.$$

Thus α is measurable.

Now let σ be any outer set covering α. Then, by Theorem 6.3.7,
$$g(\alpha) \leqslant g(\sigma),$$

whence $\qquad\qquad m^*(\alpha) = \inf g(\sigma) \geqslant g(\alpha).$

But $\qquad\qquad\qquad m^*(\alpha) \leqslant g(\alpha).$

Therefore $\qquad\qquad m(\alpha) = m^*(\alpha) = g(\alpha).$

THEOREM 7.4.3. *Any outer set is measurable and its Lebesgue measure is equal to its geometric measure.*

If σ is the union of enumerable disjoint intervals then by the complete additivity of Lebesgue measure (Theorem 7.3.5) σ is measurable and

$$m(\sigma) = \sum_{n=1}^{\infty} m(\sigma_n) = \sum_{n=1}^{\infty} g(\sigma_n) \quad \text{(by Theorem 7.4.2)}$$

$$= g(\sigma) \qquad \text{(by Definition 6.3.2).}$$

THEOREM 7.4.4. *Any inner set is measurable and its Lebesgue measure is equal to its geometric measure.*

By Theorem 7.3.1, if the inner set τ is the complement of an outer set σ with respect to an interval ω, then

$$m_*(\tau) = g(\omega) - m^*(\sigma),$$
$$m^*(\tau) = g(\omega) - m_*(\sigma).$$

But $m^*(\sigma) = m_*(\sigma) = m(\sigma)$ by Theorem 7.4.3;

whence $m_*(\tau) = m^*(\tau) = g(\omega) - m(\sigma)$

$$= g(\omega) - g(\sigma)$$

$$= g(\tau),$$

by Definition 6.6.2.

Next we shall consider 'null sets', i.e. sets of points of zero measure.

THEOREM 7.4.5. *The necessary and sufficient condition that a set α should be measurable and have measure zero is that*

$$m^*(\alpha) = 0.$$

The condition is obviously necessary for if α has measure zero then $m^*(\alpha) = m(\alpha) = 0$.

The condition is also sufficient for the relations

$$0 \leqslant m_*(\alpha) \leqslant m^*(\alpha) = 0,$$

show that $m^*(\alpha) = m_*(\alpha) = 0.$

The definition that we gave in Chapter 5 (5.2.1) is therefore completely in accord with the definition based on the theory of Lebesgue measure.

7.5. Unbounded sets

Hitherto we have restricted ourselves to *bounded* sets of points α and we have therefore been able to assert the existence of upper bracketing functions μ such that μ is an enumerable collection of intervals which cover α. In the case of *unbounded* sets bounded upper bracketing functions may not be available and we therefore rely on the method of monotony to provide a definition of measure.

DEFINITION 7.5.1. If σ_s denotes the interval

$$-s < x < s$$

an unbounded set α is measurable if the bounded sets

$$\alpha_s = \alpha\sigma_s$$

are measurable for each value of s, and the measure of α is defined to be the limit of the non-decreasing function $m(\alpha_s)$, i.e.

$$m(\alpha) = \lim_{s \to \infty} m(\alpha\sigma_s).$$

This measure $m(\alpha)$ may possibly be infinite.

7.6. Non-measurable sets

At this stage the patient reader may well be inclined to inquire if the use of both upper and lower bracketing functions and the introduction of both outer and inner measures is really necessary, since the measure of a bounded measurable set α can be defined simply as
$$m(\alpha) = \inf g(\mu)$$

for all outer sets μ covering α. It might therefore appear that only the outer measure and the upper bracketing function are necessary to define.

The answer is that such a definition would apply only to measurable sets and that measurable sets are defined only by reference to their outer and inner measures. We have not proved that all sets are measurable and hence we have relied on the bracketing process to identify the measurable sets. But the awkward question then arises—are there really any non-measurable sets, i.e. sets such that $m_*(\alpha) \neq m^*(\alpha)$?

The reply to this question falls into three parts:

(i) Examples of non-measurable sets have been constructed on the assumption that the axiom of choice of set theory is valid.

(ii) It has been proved by Solovay (1970) that the existence of non-measurable sets cannot be established if the axiom of choice is disallowed.

(iii) But the sets usually encountered in analysis are all measurable.

We shall therefore not pursue this matter further but refer the reader to the discussion in Burkill (1953), McShane (1947), and Williamson (1962).

G

7.7. Criteria for measurability

The preceding considerations suggest the desirability of constructing some simple and practical criteria of the measurability of sets. To do this we shall compare the set σ to be examined for measurability with certain outer sets α, which closely approximate to σ in the sense that the outer measure $m^*(\sigma \Delta \alpha)$ of the symmetric difference of σ and α can be made arbitrarily small. We shall thus obtain some insight into the 'structure' of a measurable set σ and we shall prove that for any tolerance $\epsilon > 0$ there exists an outer set α such that $m^*(\sigma \Delta \alpha) < \epsilon$.

The basic theorem follows at once from the definition of outer and inner measures:

THEOREM 7.7.1. *If α and β are complementary with respect to an interval ω then the necessary and sufficient condition that α and β should be measurable is that for any $\epsilon > 0$ there should exist outer sets μ and ν such that*

$$\alpha \leqslant \mu, \qquad \beta \leqslant \nu$$

and
$$g(\mu)+g(\nu) \leqslant g(\omega)+\epsilon.$$

By Definitions 7.2.1 and 7.2.2,

$$m^*(\alpha) \leqslant g(\mu), \qquad m^*(\beta) \leqslant g(\nu),$$

$$m_*(\alpha) \leqslant g(\omega-\beta) = g(\omega)-m^*(\beta).$$

Hence
$$m^*(\alpha)-m_*(\alpha) \leqslant g(\mu)-g(\omega)+m^*(\beta)$$
$$\leqslant g(\mu)+g(\nu)-g(\omega)$$

and similarly

$$m^*(\beta)-m_*(\beta) \leqslant g(\mu)+g(\nu)-g(\omega).$$

Thus the conditions of the theorem are sufficient to ensure the measurability of α and β.

Also, if α is measurable, so also is β by Theorem 7.3.1. Hence by definition for any tolerance $\epsilon > 0$ there exist outer sets μ and ν such that $\alpha \leqslant \mu$, $\beta \leqslant \nu$ and

$$m(\alpha) \geqslant g(\mu)-\tfrac{1}{2}\epsilon, \quad m(\beta) \geqslant g(\nu)-\tfrac{1}{2}\epsilon,$$

whence
$$g(\mu)+g(\nu) \leqslant m(\alpha)+m(\beta)+\epsilon = g(\omega)+\epsilon$$

by Theorem 7.3.1. Thus the conditions are necessary.

A slightly different version of this result is due to de la Vallée-Poussin.

THEOREM 7.7.2. *A necessary and sufficient condition that a bounded set σ should be measurable is that to any tolerance $\epsilon > 0$ there corresponds an inner set λ such that $\lambda \leqslant \sigma$ and $m^*(\sigma-\lambda) \leqslant \epsilon$.*

Since σ covers λ, $\sigma\lambda = \lambda$.

The union of $\eta = \sigma-\lambda$ and of λ is

$$\eta \cup \lambda = (\sigma-\lambda)+\lambda-(\sigma\lambda-\lambda) = \sigma.$$

The intersection of η and λ is

$$\eta\lambda = \sigma\lambda-\lambda = 0.$$

Hence by the union–intersection theorems

$$m^*(\eta)+m^*(\lambda) \geqslant m^*(\sigma)$$

and

$$m_*(\eta)+m_*(\lambda) \leqslant m_*(\sigma),$$

whence

$$m^*(\sigma)-m_*(\sigma) \leqslant \{m^*(\eta)-m_*(\eta)\}+\{m^*(\lambda)-m_*(\lambda)\} \leqslant m^*(\eta).$$

The condition is therefore sufficient to ensure that

$$m^*(\sigma) = m_*(\sigma).$$

The condition is also necessary. For λ is measurable, and, if σ is measurable, so also is η (by the corollary to Theorem 7.3.2) and

$$m(\sigma) = m(\lambda)+m(\eta).$$

But, by Definition 7.2.2, for every given tolerance $\epsilon > 0$, there exists an inner set λ such that $\lambda \leqslant \sigma$,

$$m(\sigma) \leqslant m(\lambda)+\epsilon,$$

whence

$$m(\eta) \leqslant \epsilon.$$

7.8. Monotone sequences of sets

By expressing Theorem 7.3.5 in terms of monotone sequences of measurable sets, instead of a series of disjoint sets, we can generalize it to apply to *any* convergent sequence of measurable sets.

THEOREM 7.8.1. *If* $\{\sigma_n\}$ *is a bounded non-decreasing sequence of measurable sets, then* $\sigma = \lim\limits_{n\to\infty} \sigma_n$ *is also measurable and*

$$m(\sigma) = \lim_{n\to\infty} m(\sigma_n).$$

Since $\{\sigma_n\}$ is bounded, there is an interval ω covering all the sets σ_n. Then $\omega - \sigma_n$ is measurable by Theorem 7.3.1, and

$$\tau_n = \sigma_n(\omega - \sigma_{n-1}) = \sigma_n - \sigma_{n-1}$$

is measurable by Theorem 7.3.2.

Now $\qquad\qquad \sigma_m \leqslant \sigma_n \quad \text{if } m < n,$

i.e. $\qquad\qquad \sigma_m \sigma_n = \sigma_m;$

whence $\qquad\qquad \tau_m \tau_n = 0.$

Thus the sets $\{\tau_n\}$ are disjoint.

But $\qquad\qquad \sigma = \sigma_1 + \sum\limits_{n=2}^{\infty} \tau_n$

and $\qquad\qquad \sigma_1 \tau_n = 0 \quad \text{for } n = 2, 3, \dots .$

Hence, by Theorem 7.3.5,

$$m(\sigma) = m(\sigma_1) + \sum_{n=2}^{\infty} m(\tau_n)$$

$$= m(\sigma_1) + \lim_{n\to\infty} \sum_{n=2}^{n} \{m(\sigma_n) - m(\sigma_{n-1})\},$$

by Theorem 7.3.2.

Therefore $\qquad\qquad m(\sigma) = \lim\limits_{n\to\infty} m(\sigma_n).$

COROLLARY. *If* $\{\sigma_n\}$ *is a bounded non-increasing sequence of measurable sets then* $\sigma = \lim\limits_{n\to\infty} \sigma_n$ *is also measurable and*

$$m(\sigma) = \lim_{n\to\infty} m(\sigma_n).$$

For we have only to apply Theorem 7.8.1 to the non-decreasing sequence $\qquad\qquad \overline{\sigma_n} = \sigma_1 - \sigma_n \quad (n = 1, 2, \dots) .$

We can now use the peak and chasm functions of Definition 3.4.2 to discuss the measurability of any bounded convergent sequence of measurable sets.

THEOREM 7.8.2. *If $\{\sigma_n\}$ is a bounded convergent sequence of measurable sets with limit σ then σ is also measurable and*

$$m(\sigma) = \lim_{n\to\infty} m(\sigma_n).$$

The associated peak and chasm sequences of Definition 3.4.2 are $\{\pi_n\}$ and $\{\chi_n\}$ where

$$\pi_n = \bigcup_{k=n}^{\infty} \sigma_k$$

and

$$\chi_n = \sigma_n \sigma_{n+1} \cdots .$$

Now

$$\pi_n \geqslant \pi_{n+1},$$

whence the sequence $\{\pi_n\}$ converges to a limit π and, by the corollary to Theorem 7.8.1,

$$m(\pi) = \lim m(\pi_n) \geqslant \limsup m(\sigma_n).$$

Similarly if each σ_n is bounded by an interval ω we have the complementary relations

$$\omega - \chi = \omega - \lim \chi_n = \lim(\omega - \chi_n)$$

and

$$m(\omega - \chi) \geqslant \lim m(\omega - \chi_n),$$

i.e.

$$m(\chi) \leqslant \lim m(\chi_n) \leqslant \liminf m(\sigma_n).$$

But

$$\chi = \pi = \sigma,$$

hence

$$m(\sigma) \leqslant \liminf m(\sigma_n) \leqslant \limsup m(\sigma_n) \leqslant m(\sigma),$$

and

$$m(\sigma) = \lim m(\sigma_n).$$

7.9. Exercises

1. A necessary and sufficient condition for a bounded set σ to be measurable is that, for any tolerance $\epsilon > 0$, there exist open sets β and α such that

$$\sigma \leqslant \beta, \qquad \beta - \sigma \leqslant \alpha$$

and

$$m(\alpha) \leqslant \epsilon.$$

2. A necessary and sufficient condition for a bounded set σ to be measurable is that, for any tolerance $\epsilon > 0$, there exists an open set β such that

$$m^*(\beta - \sigma) < \epsilon \quad \text{(Saks, Riesz)}.$$

3. A necessary and sufficient condition for a bounded set σ to be measurable is that, for any tolerance $\epsilon > 0$, there exist an *elementary* set α and two other sets η_1, η_2 such that

$$\sigma = \alpha + \eta_1 - \eta_2$$

and

$$m^*(\eta_1) < \epsilon, \quad m^*(\eta_2) < \epsilon \quad \text{(Lebesgue)}.$$

4. A necessary and sufficient condition for a bounded set σ to be measurable is that, for any tolerance $\epsilon > 0$, there exists an *elementary* set α such that

$$m^*(\delta) \leqslant \epsilon$$

where $\delta = \alpha + \sigma - 2\alpha\sigma$ is the symmetric difference of α and σ (Kolmogorov and Fomin).

5. A necessary and sufficient condition for a bounded set σ to be measurable is that, for any set τ,

$$m^*(\tau) = m^*(\tau\sigma) + m^*(\tau\sigma'),$$

where $\sigma' = 1 - \sigma$.

8 The Lebesgue integral of bounded, measurable functions

8.1. Introduction

In the preceding chapters (6 and 7) we have discussed measurable sets of points and the integration of indicators. Now, as we have indicated in § 2.4, the strategy of Lebesgue integration is to bracket the integrand $f(x)$ by a pair of functions of the form

$$\lambda(x) = \sum_{p=0}^{n-1} t_p \, \alpha_p(x),$$

$$\mu(x) = \sum_{p=0}^{n-1} t_{p+1} \alpha_p(x),$$

where $\alpha_p(x)$ is the indicator of the set of points at which

$$t_p < f(x) \leqslant t_{p+1}.$$

The problem of defining the integral of $f(x)$ is thus reduced to the problem of integrating the bracketing functions $\lambda(x)$ and $\mu(x)$ and this depends upon the problem of integrating the indicators $\alpha_p(x)$.

The problem of integration is therefore soluble, at least for bounded functions $f(x)$ and finite interval of integration, when the indicators $\alpha_p(x)$ are integrable, i.e. when the sets of points

$$t_p < f(x) \leqslant t_{p+1}$$

are measurable for all values of $t_0, t_1, ..., t_n$.

In pre-Lebesguean analysis integrals in one dimension were defined only over finite or infinite intervals, but it is one of the remarkable and important characteristics of the Lebesgue integral that it can be defined just as easily over any measurable set of points E. It is a great convenience to adopt this more general definition from the beginning.

To illustrate the concept of integration over a measurable E we may anticipate the results established later in this chapter

(§ 8.6) and say that, if $f(x)$ is integrable over interval $[a, b]$ and if $\chi(x, E)$ is the indicator of a measurable set of points E lying in this interval, then the product $f(x)\chi(x, E)$ is also integrable over $[a, b]$ and the integral of $f(x)$ over the set E is

$$\int\limits_E f(x)\,\mathrm{d}x = \int\limits_a^b f(x)\chi(x, E)\,\mathrm{d}x.$$

In uniting these two ideas—of the bracketing process and of integration over a measurable set—we shall therefore begin by studying bounded functions $f(x)$ that are defined on a bounded measurable set E, and are such that the subsets of E at which

$$t < f(x) \leqslant u$$

are measurable for all t and u.

8.2. Measurable functions

The set of points in a measurable set E at which $t < f(x) \leqslant u$ is clearly the intersection of the points in E at which $t < f(x)$ and the set of points in E at which $f(x) \leqslant u$. We therefore take as our basic definition:

DEFINITION 8.2.1. We denote by $E(f > t)$, $E(f < t)$, $E(f \geqslant t)$, $E(f \leqslant t)$, $E(f = t)$ the sets of points in a measurable set E at which $f(x) > t$, $f(x) < t$, $f(x) \geqslant t$, $f(x) \leqslant t$, $f(x) = t$, respectively.

DEFINITION 8.2.2. A function $f(x)$ is 'measurable in a measurable set E' if the set $E(f > t)$ is measurable for each value of t.

THEOREM 8.2.1. *If the set $E(f > t)$ is measurable for each value of t, so also are the sets $E(f < t)$, $E(f \geqslant t)$, $E(f \leqslant t)$, $E(f = t)$.*

The set $E(f \geqslant t)$ is the limit of the convergent sequence of measurable sets $E(f > t-1/n)$ for $n = 1, 2, \dots$. Hence, by Theorem 7.8.2 the set $E(f \geqslant t)$ is measurable.

The sets $E(f < t)$ and $E(f \geqslant t)$ are complementary with respect to the set E. Hence, by Theorem 7.3.1 the set $E(f < t)$ is measurable.

Similarly, the set $E(f \leqslant t)$ is measurable.

Finally, the set $E(f = t)$ is the intersection of the sets $E(f \geqslant t)$ and $E(f \leqslant t)$, whence, by Theorem 7.3.2, the set $E(f = t)$ is measurable.

THEOREM 8.2.2. *If $f(x)$ and $g(x)$ are each measurable in a set E, so also are the functions*

$$f(x)+c, \quad cf(x), \quad f^2(x), \quad and \quad |f(x)|,$$

where c is any real number.

The relations

$$E(f+c > t) = E(f > t-c)$$

and $$E(cf > t) = E(f > t/c) \quad (c > 0)$$

are sufficient to show that $f+c$ is measurable in E for all c, and that cf is measurable in E if c is positive.

Also the set of points $E(-f > t)$ is the same as the set $E(f < -t)$, which is measurable by Theorem 8.2.1. Hence $-f$ is measurable in E, and cf is measurable on E whether c is positive or negative.

The same theorem shows that the sets $E(f > \sqrt{t})$ and $E(f < -\sqrt{t})$ are both measurable if t is non-negative, whence their union $E(f^2 > t)$ is measurable, i.e. f^2 is measurable in E.

Similarly the set $E(|f| > t)$ is the union of the measurable sets $E(f > t)$ and $E(f < -t)$, whence $|f|$ is measurable in E.

In order to establish the measurability of $f+g$ we need the following lemma.

LEMMA. *If $f(x)$ and $g(x)$ are bounded, measurable functions in E, then the set of points $E\ (f > g)$ is measurable.*

Let $\{r_n\}$ be any enumeration of the rational numbers. Then any point ξ in E at which $f(\xi) > g(\xi)$ lies in the intersection I_n of two sets of the type $E(f > r_n)$ and $E(g < r_n)$. Hence $E(f > g)$ is the union of the enumerable, measurable sets $\{I_n\}$ and is therefore measurable.

THEOREM 8.2.3. *If f and g are measurable in E, so also are $f+g$, $af+bg$, fg, $\max(f,g)$, $\min(f,g)$.*

For $$E(f+g > t) = E(f > t-g).$$

By Theorem 8.2.2, $-g$ and $t-g$ are measurable, whence by the lemma, $f+g$ is measurable. Also by Theorem 8.2.2, af and bg are measurable, whence $af+bg$ is measurable.

Now $$fg = \tfrac{1}{4}(f+g)^2 - \tfrac{1}{4}(f-g)^2,$$

and by Theorem 8.2.2, $(f+g)^2$, $(f-g)^2$ are measurable, whence fg is measurable.

If $\phi = \max(f,g)$ and $\psi = \min(f,g)$ then

$$\phi = \tfrac{1}{2}|f-g| + \tfrac{1}{2}(f+g),$$
$$\psi = \tfrac{1}{2}(f+g) - \tfrac{1}{2}|f-g|.$$

Since $f+g, f-g, |f+g|, |f-g|$ are measurable, so also are ϕ and ψ.

Theorem 8.2.4. *If $\{f_n(x)\}$ is a sequence of functions measurable in E, then the limits*

$$\overline{\lim_{n \to \infty}} f_n(x), \quad and \quad \underline{\lim_{n \to \infty}} f_n(x).$$

are also measurable in E.

Let $$M(x) = \sup f_n(x),$$
$$L(x) = \inf f_n(x).$$

The set $E(M > t)$ is the union of the enumerable collection of measurable sets $E(f_n > t)$, and the set $E(L < t) = E(-L > -t)$ is the union of the enumerable collection of measurable sets $E(-f_n > -t)$. Hence $M(x)$ and $L(x)$ are each measurable in E.

Now let $$M_n(x) = \sup\{f_n(x), f_{n+1}(x), \ldots\},$$
and $$L_n(x) = \inf\{f_n(x), f_{n+1}(x), \ldots\}.$$
Then $$L_{n+1}(x) \geqslant L_n(x) \quad \text{and} \quad M_n(x) \geqslant M_{n+1}(x).$$
Hence $$\underline{\lim_{n \to \infty}} f_n(x) = \lim_{n \to \infty} L_n(x)$$
and $$\overline{\lim_{n \to \infty}} f_n(x) = \lim_{n \to \infty} M_n(x).$$

$L_n(x)$ and $M_n(x)$ are each measurable in E. Hence, as before, so also are $\lim L_n(x)$ and $\lim M_n(x)$, $\underline{\lim_{n \to \infty}} f_n(x)$ and $\overline{\lim_{n \to \infty}} f_n(x)$.

COROLLARY. *If $\{f_n(x)\}$ is a sequence of functions measurable in E and $f_n(x)$ converges pointwise to a limit $f(x)$ in E then $f(x)$ is measurable in E.*

For $$f(x) = \overline{\lim_{n \to \infty}} \, f_n(x) = \underline{\lim_{n \to \infty}} \, f_n(x).$$

Finally, we prove that if $f(x)$ is continuous in the interval $[a, b]$ then it is also measurable in $[a, b]$.

THEOREM 8.2.5. *The set of points in $[a, b]$ at which $f(x) > t$ is open, and is therefore an outer set, which is measurable by Theorem 7.4.3.*

All the functions of classical analysis can be constructed from the functions $f(x) = 1$ and $f(x) = x$ by the algebraic processes of addition and multiplication together with the analytic process of taking the limit of a convergent sequence. Hence all such functions are measurable.

The question of the existence of non-measurable functions, like the question of the existence of non-measurable sets of points depends upon the truth or falsehood of the axiom of choice, but we shall not explore this branch of the morbid pathology of functions.

8.3. Measure functions

DEFINITION 8.3.1. If $f(x)$ is measurable in a set E, its 'measure function' $m_E(t, f)$ is the measure of the set of points in E at which $f(x) > t$.

THEOREM 8.3.1. *The measure function $m_E(t, f)$ is a non-increasing function of t.*

For, if $s < t$, then the set of points $E(f > t)$ is covered by the set $E(f > s)$, whence

$$m_E(t, f) \leqslant m_E(s, f),$$

by Theorem 7.3.2, Corollary (ii).

COROLLARY. *If $f(x)$ is measurable in a set E which lies in a finite interval I and if $\alpha \leqslant f(x) \leqslant \beta$ in E, then the measure function*

$m_E(t,f)$ *maps the bounded domain* $\alpha \leqslant r \leqslant \beta$ *into a bounded range (since* $0 \leqslant m_E(t,f) \leqslant |I|$), *whence* $m_E(t,f)$ *is integrable by* § 2.5.

8.4. Simple functions

The functions $\lambda(x)$ and $\mu(x)$ which Lebesgue introduced to bracket a bounded function $f(x)$ belong to the class of 'simple' functions, which may be formally described in the following definitions and theorems.

DEFINITION 8.4.1. A simple function $\sigma(x)$ is one which is zero outside some finite interval (a,b) and whose range is a finite collection of distinct real numbers $s_1, s_2,..., s_m$.

THEOREM 8.4.1. *The simple function* $\sigma(x)$ *of Definition 8.4.1 can be expressed in the form*

$$\sigma(x) = \sum_{p=1}^{m} s_p\, \sigma_p,$$

where $\sigma_p = \sigma_p(x)$ *is the indicator of the points at which* $\sigma(x) = s_p$ *and where*

$$\sum_{p=1}^{m} \sigma_p = 1.$$

For if x is any prescribed point, then $\sigma(x)$ has one and only one of the values $s_1, s_2,..., s_m$, say s_k, and

$$\sum_{p=1}^{m} s_p\, \sigma_p(x) = s_k = \sigma(x).$$

To define the integral of the simple function $\sigma(x)$ we could follow the same methods as those we used to define the integral of an indicator function (Chapter 7), but a few moments reflection should satisfy the reader that the final result would be expressed by

DEFINITION 8.4.2. If the simple function $\sigma(x)$ is measurable, then its integral is

$$\int \sigma(x)\, \mathrm{d}x = \sum_{p=1}^{m} s_p \int \sigma_p(x)\, \mathrm{d}x = \sum_{p=1}^{m} s_p\, m_p,$$

where m_p is the measure of the set of points σ_p at which $\sigma(x) = s_p$.

To justify this definition we prove

THEOREM 8.4.2. *The integral $\int \sigma(x)\,dx$ of a simple function $\sigma(x)$ is a positive, linear functional on the space of simple functions.*

For, if $\sigma(x)$ is non-negative, then

$$s_p \geqslant 0 \quad \text{for } p = 1, 2, ..., m,$$

whence

$$\int \sigma(x)\,dx \geqslant 0.$$

Now let the simple functions $\sigma(x)$ and $\tau(x)$ have representations

$$\sigma(x) = \sum_p s_p \alpha_p, \qquad \tau(x) = \sum_q t_q \beta_q,$$

with

$$\sum \alpha_p = 1, \qquad \sum \beta_q = 1.$$

Then

$$\sigma(x) + \tau(x) = \sum s_p \alpha_p + \sum t_q \beta_q$$
$$= \sum (s_p + t_q)\alpha_p \beta_q.$$

Also

$$\sum \alpha_p \beta_q = \sum \alpha_p \sum \beta_q = 1.$$

The set of values $(s_p + t_q)$ may not all be distinct, but they can be grouped into a finite collection of distinct values $u_1, u_2, ...$ and then $\sum \alpha_p \beta_q$ summed over all p and q for which $s_p + t_q = u_k$ will be the indicator $\chi(x, u_k)$ of the points at which $\sigma(x) = u_k$. Hence the sum of two simple functions is a simple function.

Now, by Definition 8.4.2,

$$\int \{\sigma(x) + \tau(x)\}\,dx = \sum_k u_k \int \chi(x, u_k)\,dx$$
$$= \sum_{p,q} (s_p + t_q) \int \alpha_p \beta_q\,dx.$$

The set of points at which $\sigma(x) = s_p$ is the union of the finite collection of disjoint sets at which $\sigma(x) = s_p$ and $\tau(x) = t_q$ for $q = 1, 2, ...$.

Hence, by Theorem 7.3.2, Corollary (i),

$$\int \alpha_p\,dx = \sum_q \int \alpha_p \beta_q\,dx.$$

Similarly

$$\int \beta_q\,dx = \sum_p \int \alpha_p \beta_q\,dx.$$

Therefore

$$\int \{\sigma(x) + \tau(x)\}\,dx = \sum_{p,q} s_p \int \alpha_p \beta_q\,dx + \sum_{p,q} t_q \int \alpha_p \beta_q\,dx$$
$$= \sum_p s_p \int \alpha_p\,dx + \sum_q t_q \int \beta_q\,dx$$
$$= \int \sigma(x)\,dx + \int \tau(x)\,dx.$$

Similarly we can show that, if c and k are any real numbers, $c\sigma(x)$, $k\tau(x)$ and $c\sigma(x)+k\tau(x)$ are simple functions, and that

$$\int \{c\sigma(x)+k\tau(x)\}\,\mathrm{d}x = c\int \sigma(x)\,\mathrm{d}x + k\int \tau(x)\,\mathrm{d}x.$$

Thus the integral of a simple function is a linear functional.

8.5. Lebesgue bracketing functions

The following definition is effectively the same as that given by Lebesgue.

DEFINITION 8.5.1. If E is a measurable set of points and if $f(x)$ is a bounded function and its range, $A < f(x) \leqslant B$, is divided at the finite number of points $t_0, t_1,..., t_n$ so that

$$A = t_0 < t_1 < t_2 < ... < t_n = B,$$

then

$$\lambda(x) = \sum_{p=0}^{n-1} t_p\{\alpha_E(x, t_p) - \alpha_E(x, t_{p+1})\}$$

and

$$\mu(x) = \sum_{p=0}^{n-1} t_{p+1}\{\alpha_E(x, t_p) - \alpha_E(x, t_{p+1})\}$$

are lower and upper Lebesgue bracketing functions for $f(x)$, $\alpha_E(x, t_p)$ being the indicator of the points in E at which $f(x) > t_p$.

THEOREM 8.5.1. *The Lebesgue bracketing functions of a bounded measurable function $f(x)$ defined on a bounded, measurable set E are* uniform *approximations to $f(x)$ on E, and*

$$\sup \int_a^b \lambda(x)\,\mathrm{d}x = \inf \int_a^b \mu(x)\,\mathrm{d}x.$$

Let $\epsilon = \max(t_{p+1} - t_p)$ for $p = 0, 1, 2,..., n-1$. Then, if $x \in E$,

$$\lambda(x) \leqslant f(x) \leqslant \mu(x),$$

and

$$\mu(x) - \lambda(x) \leqslant \epsilon.$$

Hence the functions $\lambda(x)$ and $\mu(x)$ approximate $f(x)$ uniformly in E.

The bracketing functions are simple functions. Hence they are integrable and, by Theorem 8.4.2,

$$\int_a^b \mu(x)\,\mathrm{d}x - \int_a^b \lambda(x)\,\mathrm{d}x = \int_a^b \{\mu(x)-\lambda(x)\}\,\mathrm{d}x \leqslant \epsilon(b-a).$$

Now $\lambda(x) \leqslant B$ and $\mu(x) \geqslant A$. Hence if we consider all the Lebesgue functions of a bounded measurable function $f(x)$, then the bounds

$$\sup \int_a^b \lambda(x)\,\mathrm{d}x \quad \text{and} \quad \inf \int_a^b \mu(x)\,\mathrm{d}x$$

both exist, and

$$\inf \int_a^b \mu(x)\,\mathrm{d}x - \sup \int_a^b \lambda(x)\,\mathrm{d}x \leqslant \epsilon(b-a).$$

Since this is true for all $\epsilon > 0$ it follows that

$$\inf \int_a^b \mu(x)\,\mathrm{d}x = \sup \int_a^b \lambda(x)\,\mathrm{d}x.$$

8.6. The Lebesgue–Young integral

We are at last in a position to define the Lebesgue integral of any bounded, measurable function $f(x)$ over a bounded measurable set E. In accordance with our general bracketing principle, which has been our main guide from Chapter 1 onwards, we now give

DEFINITION 8.6.1. If $\lambda(x)$ and $\mu(x)$ are Lebesgue bracketing functions of a bounded, measurable function $f(x)$, for a bounded, measurable set E, then the Lebesgue integral of $f(x)$ over E is defined to be

$$\int_E f(x)\,\mathrm{d}x = \sup \int_a^b \lambda(x)\,\mathrm{d}x = \inf \int_a^b \mu(x)\,\mathrm{d}x.$$

We shall justify this definition in § 8.7 by showing that the Lebesgue integral is a positive, linear, continuous functional and by proving that almost everywhere in (a,b) the derivative

$$\frac{\mathrm{d}}{\mathrm{d}x} \int_a^x f(t)\,\mathrm{d}t$$

exists and equals $f(x)$.

The Lebesgue integral of $f(x)$ can be expressed as the integral of the measure function $m_E(t,f)$ in the form given by

THEOREM 8.6.1. *If $A \leqslant f(x) \leqslant B$, for all $x \in E$ then*

$$\int_E f(x)\, \mathrm{d}x = Am(E) + \int_A^B m_E(t,f)\, \mathrm{d}t.$$

Let $\qquad m_k = m_E(t_k, f) = \int \alpha(x, t_n, f)\, \mathrm{d}x.$

Then $\qquad m_0 = m(E) \quad \text{and} \quad m_n = 0.$

Also $\qquad t_0 = A < t_1.$

$$\int_E \lambda(x)\, \mathrm{d}x = t_0(m_0 - m_1) + t_1(m_1 - m_2) + \dots + t_{n-1}(m_{n-1} - m_n)$$

$$= m_0 t_0 + m_1(t_1 - t_0) + \dots + m_{n-1}(t_{n-1} - t_{n-2}).$$

Hence, by Theorem 2.5.1 on the integration of monotone functions,

$$\int_E \lambda(x)\, \mathrm{d}x \leqslant Am(E) + \int_A^B m_E(t, f)\, \mathrm{d}t.$$

Similarly

$$\int_E \mu(x)\, \mathrm{d}x = t_1(m_0 - m_1) + t_2(m_1 - m_2) + \dots + t_n(m_{n-1} - m_n)$$

$$= m_0 t_1 + m_1(t_2 - t_1) + \dots + m_{n-1}(t_n - t_{n-1})$$

$$\geqslant Am(E) + \int_A^B m_E(t, f)\, \mathrm{d}t.$$

Therefore, by Definition 8.6.1,

$$\int_E f(x)\, \mathrm{d}x = Am(E) + \int_A^B m_E(t, f)\, \mathrm{d}t.$$

This representation of the Lebesgue integral in terms of the measure function $m_E(t, f)$ was the independent discovery of Lebesgue and of W. H. Young (1905). While the independent variable x varies over the set E, the dependent variable $t = f(x)$ varies over the range (A, B), so that in the definition of the Lebesgue integral the roles of the independent and dependent variables are interchanged. This inversion of the roles of the independent and dependent variables is well described by G. H. Hardy as a 'dramatic' transformation.

COROLLARY. (i) *We note in particular that if $f(x)$ is bounded and measurable in the closed interval $[a, b]$, then it is also bounded and measurable in the open interval (a, b) and the half-open intervals $(a, b]$, $[a, b)$. Each of these intervals has the same measure $b-a$, and the measure function $m_E(t, f)$ has the same value in each case. Therefore we can use the same symbol*

$$\int_a^b f(x) \, \mathrm{d}x$$

for the integral of $f(x)$ over $[a, b]$, (a, b), $[a, b)$ or $(a, b]$.

(ii) *It is almost trivial to note that if $f(x)$ is bounded and measurable in $[a, b]$ then we can similarly define the Lebesgue integral*

$$\int_\alpha^\beta f(x) \, \mathrm{d}x$$

for any interval $[\alpha, \beta]$ covered by $[a, b]$.

Note. It is convenient to define $\int_b^a f(x) \, \mathrm{d}x$ for $a < b$ to mean

$$-\int_a^b f(x) \, \mathrm{d}x.$$

There is a companion theorem to 8.6.1 which can be proved in exactly the same way and which yields

THEOREM 8.6.2. *If $f(x)$ is a bounded, measurable function and $A < f(x) \leqslant B$ in the measurable set E and $\mu_E(t, f)$ is the measure of the set of points in E at which $f(x) \leqslant t$,*

then
$$\int_E f(x) \, \mathrm{d}x = Bm(E) - \int_A^B \mu_E(t, f) \, \mathrm{d}t.$$

We can also deduce this result from Theorem 8.6.1 by introducing $\beta(x, t)$ as the indicator of the points in E at which $f(x) \leqslant t$.

Then
$$\alpha(x, t) + \beta(x, t) = \chi_E(x),$$

$$\mu_E(t, f) = \int_a^b \beta(x, t) \, \mathrm{d}x = m(E) - m_E(t, f)$$

114   *Lebesgue integral of bounded, measurable functions*

and

$$Am(E)+\int_A^B m_E(t,f)\,dt = Am(E)+\int_A^B \{m(E)-\mu_E(t,f)\}\,dt$$

$$= Bm(E)-\int_A^B \mu_E(t,f)\,dt$$

by the additive property of the integrals of monotone function (§ 2.7, Exercise 4).

8.7. The Lebesgue integral as a positive, linear, continuous functional

Throughout this section $f(x)$ denotes a bounded, measurable function and E a bounded, measurable set of points.

To justify the definition (8.6.1) of the Lebesgue integral of $f(x)$ over E we must show that $\int_E f(x)\,dx$ is a positive, linear continuous functional of $f(x)$ with the Lebesgue norm. We shall in fact establish the theorem of bounded convergence (8.7.7) anticipated in § 2.2.

THEOREM 8.7.1 (the mean value theorem). *If $A < f(x) \leqslant B$, then $Am(E) < \int_E f(x)\,dx \leqslant Bm(E)$.*

For
$$\lambda(x) \geqslant A\{\alpha_E(x,A)-\alpha_E(x,B)\}$$
$$\mu(x) \leqslant B\{\alpha_E(x,A)-\alpha_E(x,B)\},$$

whence
$$\int_E \lambda(x)\,dx \geqslant Am(E)$$

and
$$\int_E \mu(x)\,dx \leqslant Bm(E),$$

where
$$m(E) = \int_E \{\alpha_E(x,t_0)-\alpha_E(x,t_n)\}\,dx,$$

i.e. the measure of the set of points in E at which $A < f(x) \leqslant B$. i.e. the measure of the set E.

COROLLARY. (i) *If $A = B = C$ then*
$$\int_E f(x)\,dx = Cm(E).$$

If E is the interval (a, b), and $f(x) = 1$ then

$$\int_a^b f(x) \, \mathrm{d}x = (b-a).$$

Thus the Lebesgue integral satisfies the Lebesgue normalizing condition of § 2.2.

(ii) *If $A \geqslant 0$, then $\int_E f(x) \, \mathrm{d}x \geqslant 0$.*

Hence the Lebesgue integral is a positive functional on the space of bounded measurable functions.

THEOREM 8.7.2 (the addition theorem for sets). *If the bounded set E is the union of a finite or enumerable collection of disjoint measurable sets $\{E_n\}$ and $f(x)$ is bounded and measurable on E, then*

$$\int_E f(x) \, \mathrm{d}x = \sum_{n=1}^{\infty} \int_{E_n} f(x) \, \mathrm{d}x.$$

If $\lambda(x)$ and $\mu(x)$ are Lebesgue bracketing functions for $f(x)$ and the set E, and if $\chi_p = \chi(x, E_p)$ is the indicator of the set E_p then by Definition 8.5.1, $\chi_p \lambda$ and $\chi_p \mu$ are Lebesgue bracketing functions for $f(x)$ and the set E_p. For, if $x \in E_p$, then

$$\chi_p \lambda \leqslant f(x) \leqslant \chi_p \mu.$$

Similarly $\qquad \sum_{p=1}^n \chi_p \lambda \quad \text{and} \quad \sum_{p=1}^n \chi_p \mu$

are Lebesgue bracketing functions for $f(x)$ and the union U_n of the sets $E_1, E_2,..., E_n$.

All these bracketing functions are simple, whence by Theorem 8.4.2 on sums of simple functions,

$$\sum \int \chi_p \lambda \, \mathrm{d}x \leqslant \int_{U_n} f(x) \, \mathrm{d}x \leqslant \sum \int \chi_p \mu \, \mathrm{d}x.$$

Also $\qquad \int \chi_p \lambda \, \mathrm{d}x \leqslant \int_{E_p} f(x) \, \mathrm{d}x \leqslant \int \chi_p \mu \, \mathrm{d}x.$

We can choose the functions λ and μ so that

$$\int_E \chi_p \mu \, \mathrm{d}x - \int_E \chi_p \lambda \, \mathrm{d}x \leqslant \int \mu \, \mathrm{d}x - \int \lambda \, \mathrm{d}x \leqslant \epsilon/n.$$

Hence, on passing to the limit,

$$\int\limits_{U_n} f(x)\ \mathrm{d}x = \sum_{p=1}^{n} \int\limits_{E_p} f(z)\ \mathrm{d}x.$$

Now let V_n be the union of the sets $E_{n+1}, E_{n+2},\dots$. Then, by the result just established,

$$\int\limits_{E} f(x)\ \mathrm{d}x = \int\limits_{U_n} f(x)\ \mathrm{d}x + \int\limits_{V_n} f(x)\ \mathrm{d}x.$$

But, by the mean value theorem (8.7.1),

$$Am(V_n) \leqslant \int\limits_{V_n} f(x)\ \mathrm{d}x \leqslant Bm(V_n),$$

if $A \leqslant f(x) \leqslant B$ in E. Also

$$m(V_n) = m(E) - m(U_n) \to 0 \quad \text{as } n \to \infty.$$

Therefore

$$\int\limits_{E} f(x)\ \mathrm{d}x = \lim_{n\to\infty} \int\limits_{U_n} f(x)\ \mathrm{d}x = \sum_{n=1}^{\infty} \int\limits_{E_p} f(x)\ \mathrm{d}x.$$

COROLLARY. (i) *By the corollary to Theorem* 8.6.1 *it follows that if $f(x)$ is bounded and measurable on $[a,b]$ and if $a < c < b$, then*

$$\int\limits_{a}^{b} f(x)\ \mathrm{d}x = \int\limits_{a}^{c} f(x)\ \mathrm{d}x + \int\limits_{c}^{b} f(x)\ \mathrm{d}x.$$

(ii) *If $f(x) = 0$ p.p. in E, then $\int\limits_{E} f(x) = 0$.*

For if $f(x) = 0$ in N and $f(x) \neq 0$ in M, then $m(M) = 0$ and

$$\int\limits_{E} f(x)\ \mathrm{d}x = \int\limits_{N} f(x)\ \mathrm{d}x + \int\limits_{M} f(x)\ \mathrm{d}x = 0.$$

We can now prove the following extension of the mean value theorem (8.7.1).

THEOREM 8.7.3. *If $f(x)$ and $\phi(x)$ are each bounded and measurable on the set E, and if $\phi(x) \geqslant f(x)$ on E then*

$$\int\limits_{E} \phi(x)\ \mathrm{d}x \geqslant \int\limits_{E} f(x)\ \mathrm{d}x.$$

If $\lambda(x)$ is the lower Lebesgue bracketing function for $f(x)$ as given in Definition 8.5.1, then $\lambda(x)$ is also a lower bracketing function for $\phi(x)$. For we have only to choose A and B so that

$$A < f(x) \leqslant \phi(x) \leqslant B.$$

Hence
$$\int_E \phi(x)\,\mathrm{d}x \geqslant \int_E \lambda(x)\,\mathrm{d}x,$$

and
$$\int_E \phi(x)\,\mathrm{d}x \geqslant \sup \int_E \lambda(x)\,\mathrm{d}x = \int_E f(x)\,\mathrm{d}x.$$

The mean value theorem enables us to infer upper and lower bounds for the integral of $f(x)$ from upper and lower bounds for $f(x)$. It is obvious that there cannot be any exact converse of this theorem, but there are two theorems that are converses of Corollary (ii) of Theorem 8.7.2.

THEOREM 8.7.4 (the null integral theorem)

(i) *If $f(x)$ is bounded, measurable and non-negative in a measurable set E, and $\int_E f(x)\,\mathrm{d}x = 0$, then $f(x) = 0$ p.p. in E.*

Let E_n be the set of points in E at which $f(x) > 1/n$. Then E_n is measurable and

$$0 \leqslant \frac{1}{n} m(E_n) \leqslant \int_{E_n} f(x)\,\mathrm{d}x \leqslant \int_E f(x)\,\mathrm{d}x = 0$$

(by Theorem 8.7.1) whence
$$m(E_n) = 0.$$

Now the set of points P on E at which $f(x) > 0$ is the union of the enumerable sets $\{E_n\}$ $(n = 1, 2,...)$. Hence the measure of the set P is zero, i.e. $f(x) = 0$ p.p. in E.

(ii) *If $f(x)$ is bounded and measurable in an interval I and $\int_J f(x)\,\mathrm{d}x = 0$ for any interval J covered by I then $f(x) = 0$ p.p. in I.*

Let $f(x) > 0$ in a set P in I. Since P is measurable it covers an inner set Q, and $f(x) > 0$ in Q. The complement, $I - Q$, is an outer set, i.e. an enumerable collection of intervals E_n. Hence

$$\int_Q f(x)\,\mathrm{d}x = \int_I f(x)\,\mathrm{d}x - \sum_{n=1}^{\infty} \int_{E_n} f(x)\,\mathrm{d}x = 0.$$

Therefore, by part (i), $f(x) = 0$ p.p. in Q, which is a contradiction, whence the result follows.

So far we have followed the concise exposition of de la Vallée-Poussin rather closely, but we can remove a certain artificiality from his proof of the following addition theorems by using the properties of simple functions.

THEOREM 8.7.5 (the addition theorem for functions). *If the functions f_1 and f_2 are bounded and measurable in E so also is f_1+f_2 and*

$$\int_E (f_1+f_2)\, \mathrm{d}x = \int_E f_1\, \mathrm{d}x + \int_E f_2\, \mathrm{d}x.$$

If λ_1, μ_1, and λ_2, μ_2 are lower and upper Lebesgue bracketing functions for f_1 and f_2 respectively, then $\lambda_1+\lambda_2$ and $\mu_1+\mu_2$ are bracketing functions for f_1+f_2, although they are not Lebesgue bracketing functions. Nevertheless they are simple functions, whence, by Theorem 8.4.2,

$$\int_E \lambda_1\, \mathrm{d}x + \int_E \lambda_2\, \mathrm{d}x = \int_E (\lambda_1+\lambda_2)\, \mathrm{d}x.$$

But, by Theorem 8.7.3,

$$\int_E (\lambda_1+\lambda_2)\, \mathrm{d}x \leqslant \int_E (f_1+f_2)\, \mathrm{d}x\,;$$

whence

$$\int_E \lambda_1\, \mathrm{d}x + \int_E \lambda_2\, \mathrm{d}x \leqslant \int_E (f_1+f_2)\, \mathrm{d}x,$$

and, similarly,

$$\int_E \mu_1\, \mathrm{d}x + \int_E \mu_2\, \mathrm{d}x \geqslant \int_E (f_1+f_2)\, \mathrm{d}x.$$

Now we can choose λ_1, λ_2, μ_1, μ_2 so that, given any tolerance $\epsilon > 0$,

$$\int_E \mu_1\, \mathrm{d}x - \epsilon < \int f_1\, \mathrm{d}x < \int \lambda_1\, \mathrm{d}x + \epsilon,$$

$$\int_E \mu_2\, \mathrm{d}x - \epsilon < \int f_2\, \mathrm{d}x < \int \lambda_2\, \mathrm{d}x + \epsilon.$$

Since this is true for all $\epsilon > 0$,

$$\int_E (f_1+f_2)\, \mathrm{d}x = \int_E f_1\, \mathrm{d}x + \int_E f_2\, \mathrm{d}x.$$

We can now show that the Lebesgue integral is a linear functional.

THEOREM 8.7.6. *If $f(x)$ and $g(x)$ are bounded and measurable on a measurable set E, and if a, b are any real numbers, then*

$$\int_E (af+bg)\,\mathrm{d}x = a\int_E f\,\mathrm{d}x + b\int_E g\,\mathrm{d}x.$$

If $a \geqslant 0$, and if λ, μ are lower and upper Lebesgue bracketing functions to $f(x)$, then $a\lambda$, $a\mu$ are lower and upper Lebesgue bracketing functions to $af(x)$. Hence

$$\int_E af\,\mathrm{d}x = a\int_E f\,\mathrm{d}x \quad (a \geqslant 0).$$

Also, by Theorem 8.7.5,

$$\int_E (-af)\,\mathrm{d}x + \int_E (af)\,\mathrm{d}x = 0,$$

whence $\quad \int_E (-af)\,\mathrm{d}x = -\int af\,\mathrm{d}x = -a\int f\,\mathrm{d}x.$

Therefore $\quad \int_E af\,\mathrm{d}x = a\int f\,\mathrm{d}x,$

whether a is positive or negative (or zero).

Finally, by the same Theorem 8.7.5,

$$\int_E (af+bg)\,\mathrm{d}x = \int_E af\,\mathrm{d}x + \int_E bg\,\mathrm{d}x = a\int_E f\,\mathrm{d}x + b\int_E g\,\mathrm{d}x.$$

COROLLARY. *By induction it follows that if $f_1, f_2, ..., f_n$ is any finite collection of functions, each bounded and measurable on a measurable set E, and if $c_1, c_2, ..., c_n$ are any real numbers, then*

$$\int_E \sum_{p=1}^{n} c_p f_p \,\mathrm{d}x = \sum_{p=1}^{n} c_p \int f_p \,\mathrm{d}x.$$

The extension of this result to an enumerable collection of functions is one of the most powerful and attractive features of the Lebesgue theory. We shall begin by proving the 'theorem of bounded convergence', which applies to a sequence $\{f_n(x)\}$ $(n = 1, 2, ...)$, such that, at each point x of a bounded and measurable set E, $|f_n(x)|$ is less than a constant K independent of n, and $f_n(x)$ converges to a limit function $f(x)$.

THEOREM 8.7.7. *If the sequence $\{f_n(x)\}$ converges boundedly to $f(x)$ in a bounded and measurable set E, then*

$$\int_E f_n(x)\,\mathrm{d}x \to \int_E f(x)\,\mathrm{d}x \quad as\ n\to\infty.$$

Choose any positive number ϵ. Let E_1 be the set of points at which

$$|f(x)-f_n(x)| < \epsilon,$$

for all n and let E_{k+1} be the set of points at which

$$|f(x)-f_k(x)| \geqslant \epsilon$$

and

$$|f(x)-f_n(x)| < \epsilon \quad \text{for } n \geqslant k+1.$$

Then the sets $\{E_k\}$ ($k = 1, 2,...$) are disjoint and their union is E. Hence, by Theorem 8.7.2,

$$\int_E f_n(x)\,\mathrm{d}x = \sum_{k=1}^{\infty} \int_{E_k} f_n(x)\,\mathrm{d}x.$$

By the addition theorems (8.7.2) and (8.7.5),

$$\int_E f(x)\,\mathrm{d}x - \int_E f_n(x)\,\mathrm{d}x = \int_E \{f(x)-f_n(x)\}\,\mathrm{d}x$$

$$= \sum_{k=1}^{n} \int_{E_k} \{f(x)-f_n(x)\}\,\mathrm{d}x + \int_{F_n} \{f(x)-f_n(x)\}\,\mathrm{d}x$$

where $F_n = E - E_1 - E_2 - ... - E_n$.

By the mean value theorem (8.7.1),

$$\left| \int_{E_k} \{f(x)-f_n(x)\}\,\mathrm{d}x \right| < \epsilon m(E_k) \quad \text{if } n \geqslant k+1,$$

whence

$$\left| \sum_{k=1}^{n} \int_{E_k} \{f(x)-f_n(x)\}\,\mathrm{d}x \right| < \epsilon \sum_{k=1}^{n} m(E_k) < \epsilon m(E).$$

Since $f_n(x)$ converges boundedly to $f(x)$, there is a constant B such that $|f(x)-f_n(x)| < B$ for all n and all x in E. Hence

$$\left| \int_{F_n} \{f(x)-f_n(x)\}\,\mathrm{d}x \right| < Bm(F_n)$$

and

$$\left| \int_E f(x)\,\mathrm{d}x - \int_E f_n(x)\,\mathrm{d}x \right| < \epsilon m(E) + Bm(F_n).$$

Now

$$E_1 + E_2 + ... + E_n \to E \quad \text{as } n\to\infty,$$

whence $\qquad\qquad m(F_n) \to 0 \quad$ as $n \to \infty.$

Therefore

$$\limsup_{n \to \infty} \left| \int_E f(x)\, dx - \int_E f_n(x)\, dx \right| < \epsilon m(E),$$

i.e. $\qquad\qquad \int_E f_n(x)\, dx \to \int_E f(x)\, dx \quad$ as $n \to \infty.$

8.8. The differentiability of the indefinite Lebesgue integral

To complete the discussion of the Lebesgue integral of a bounded measurable function $f(x)$ over a bounded measurable set of points E we must return to the special and familiar case when E is an interval $[a, t]$ and discuss the properties of the indefinite integral,

$$\phi(t) = \int_a^t f(x)\, dx,$$

regarded as a function of t in some interval $[a, b]$. We shall prove that almost everywhere in $[a, b]$, $\phi(t)$ possesses a derivative $\phi'(t)$ equal to $f(t)$.

This in fact is the best possible result that we can hope to prove for if $\chi(x, E)$ is the indicator of Cantor's ternary set E (§ 5.3), then

$$m(t) \equiv \int_0^t \chi(x, E)\, dx = 0, \quad \text{if } 0 \leqslant t \leqslant 1;$$

whence $\qquad\qquad m'(t) = 0,$

but $\qquad\qquad \chi(t, E) = 1$

if $t \in E$, i.e. at a set of points of measure zero.

We therefore begin by considering the simplest Lebesgue integral, i.e. the measure of a measurable set of points.

THEOREM 8.8.1. *If $\chi(x)$ is the indicator of a set of points E and $m(x) = \int_a^x \chi(t)\, dt$, then, almost everywhere, the measure function $m(x)$ has a derivative $m'(x)$ equal to $\chi(x)$.*

Since
$$m(y)-m(x) = \int_x^y \chi(t)\,dt \geqslant 0,$$
if $y > x$, therefore
$$0 \leqslant m(y)-m(x) \leqslant y-x,$$
whence $m(x)$ is continuous and is a non-decreasing function of x. Therefore, by Theorem 5.7.5, $m(x)$ possesses a derivative $m'(x)$, almost everywhere and $0 \leqslant m'(x) \leqslant 1$. It only remains to prove that, almost everywhere $m'(x) = \chi(x)$. Let
$$m_n(x) = n\{m(x+1/n)-m(x)\} \quad (n = 1, 2,...).$$
Then $m_n(x)$ converges boundedly to $m'(x)$ almost everywhere in any interval $[a,b]$. Therefore by the theorem of bounded convergence (8.7.7), if $a \leqslant c \leqslant b$
$$\int_a^c m_n(x)\,dx \to \int_a^c m'(x)\,dx \quad \text{as } n \to \infty.$$

But
$$\int_a^c m_n(x)\,dx = n\int_a^c m(x+1/n)\,dx - n\int_a^c m(x)\,dx$$
$$= n\int_c^{c+1/n} m(x)\,dx - n\int_a^{a+1/n} m(x)\,dx,$$
for all $n > (b-c)^{-1}$.

Now, by the mean value theorem (8.7.1), if
$$A = \inf m(x) \quad \text{and} \quad B = \sup m(x)$$
for $a \leqslant x \leqslant a+1/n$, then
$$A/n \leqslant \int_a^{a+1/n} m(x)\,dx \leqslant B/n.$$

But, since $m(x)$ is continuous,
$$A \to m(a) \quad \text{and} \quad B \to m(a) \quad \text{as } n \to \infty.$$

Therefore
$$n\int_a^{a+1/n} m(x)\,dx \to m(a) \quad \text{as } n \to \infty.$$

Similarly
$$n\int_c^{c+1/n} m(x)\,dx \to m(c) \quad \text{as } n \to \infty.$$

Therefore

$$\int\limits_a^c m'(x)\,\mathrm{d}x = m(c) - m(a) = \int\limits_a^c \chi(x)\,\mathrm{d}x,$$

for any interval $[a, c]$.

Hence, by Theorem 8.7.4,

$$m'(x) = \chi(x) \quad \text{p.p. in } [a, b].$$

A point ξ is said to be a point of metric density of the set E with indicator $\chi(x)$ if the measure function

$$m(x) = \int\limits_a^x \chi(t)\,\mathrm{d}t$$

has a derivative $m'(\xi)$ at the point ξ. Hence we have proved that almost all points are points of metric density and that the value of the metric density is 1 if $x \in E$ or 0 if $x \notin E$.

We can now examine the differentiability of the Lebesgue integral of a bounded, measurable function $f(x)$ over a bounded interval $[a, b]$.

THEOREM 8.8.2. *Almost everywhere in $[a, b]$ the integral $\phi(x) = \int\limits_a^x f(t)\,\mathrm{d}t \ (a < x < b)$ possesses a derivative equal to $f(x)$.*

If $\alpha(x, t)$ is the indicator of the set of points in $[a, b]$ at which $f(x) > t$ then the measure function

$$m(x) = \int\limits_a^x \alpha(s, t)\,\mathrm{d}s$$

has, almost everywhere in $[a, b]$, a derivative $m'(x)$ equal to $\alpha(x, t)$ by Theorem 8.8.1.

Hence, if $\lambda(x)$ and $\mu(x)$ are the Lebesgue bracketing functions of Definition 8.5.1, the integrals

$$\int\limits_a^x \lambda(s)\,\mathrm{d}s \quad \text{and} \quad \int\limits_a^x \mu(s)\,\mathrm{d}s$$

have, almost everywhere in $[a, b]$ derivatives which are respectively equal to $\lambda(x)$ and $\mu(x)$.

But, by Theorem 8.7.3,

$$\int\limits_{x}^{x+h} \lambda(s)\, \mathrm{d}s \leqslant \int\limits_{x}^{x+h} f(s)\, \mathrm{d}s \leqslant \int\limits_{x}^{x+h} \mu(s)\, \mathrm{d}s.$$

Let

$$f_h(x) = \frac{1}{h} \int\limits_{x}^{x+h} f(s)\, \mathrm{d}s.$$

Then, at almost every point in $[a, b]$,

$$\lambda(x) \leqslant \liminf_{h \to 0} f_h(x) \leqslant \limsup_{h \to 0} f_h(x) \leqslant \mu(x).$$

But we can choose the bracketing functions so that

$$\lambda(x) \leqslant f(x) \leqslant \mu(x),$$

and $\mu(x) - \lambda(x) < 1/n$ for each tolerance $1/n > 0$. Hence, for each positive integer n,

$$f(x) - 1/n \leqslant \liminf_{h \to 0} f_h(x) \leqslant \limsup_{h \to 0} f_h(x) \leqslant f(x) + 1/n,$$

except in a set of measure zero. The union of an enumerable collection of sets of measure zero has measure zero. Hence, almost everywhere in $[a, b]$,

$$\limsup_{h \to 0} f_h(x) = \liminf_{h \to 0} f_h(x) = f(x),$$

i.e.

$$D^+\phi(x) = f(x) = D_+\phi(x).$$

Similarly we can prove that

$$D^-\phi(x) = f(x) = D_-\phi(x),$$

whence the integral $\phi(x)$ possesses a derivative equal to $f(x)$ almost everywhere in $[a, b]$.

8.9. Exercises

1. If the set $E(f \geqslant t)$ is measurable, deduce the measurability of $E(f > t)$.

2. If $f(x)$ is measurable and Ω is an outer set of points, show that the set of points E for which $f(x) \in \Omega$ is measurable.

3. If $f(x)$ is continuous and $g(x)$ is measurable, prove that $f[g(x)]$ is measurable.

4. Evaluate the indicator $\beta(x, \tau)$ of the indicator $\alpha(x, \sigma)$ of a measurable function $f(x)$ and evaluate

$$\int\limits_{0}^{1} \beta(x, \tau)\, \mathrm{d}x.$$

5. If $f(x)$ is bounded and measurable in $[a, b]$, prove that for each tolerance $\epsilon > 0$ there exists a step function $g(x)$ such that

$$\int_a^b |f(x) - g(x)|\, \mathrm{d}x \leqslant \epsilon.$$

6. If the sequence of bounded, measurable functions $\{f_n(x)\}$ converges to a limit function $f(x)$ at all points of a measurable set E, prove that for each tolerance $\epsilon > 0$ there exists a measurable set F such that $F \subset E$, $m(F) > m(E) - \epsilon$, and $f_n(x) \to f(x)$ uniformly in F as $n \to \infty$.

9 The Lebesgue integral of summable functions

9.1. Introduction

In the preceding chapter we have given the analytic construction of the integral of a bounded, measurable function over a bounded measurable set by the methods of Lebesgue, and following in the path of de la Vallée-Poussin we have shown that the Lebesgue integral is a positive, linear, 'continuous' functional. Lastly we have shown that the indefinite integral

$$\phi(x) = \int_a^x f(s) \, ds$$

has almost everywhere a derivative equal to $f(x)$.

So far as bounded functions and bounded intervals are concerned it does appear that the Lebesgue integral furnishes the most general concept of an integral, but it is clearly desirable to extend this concept to unbounded functions and to unbounded intervals. Here the methods of Lebesgue do not meet with such complete success. In fact the theory applies only to the class of functions now called 'summable',† and for further generalizations it is necessary to employ the more powerful methods of Denjoy, Perron, Ward, and Henstock.

9.2. Summable functions

We have proved in Theorems 8.6.1 and 8.6.2 that the Lebesgue integral of a function $f(x)$, bounded and measurable in a bounded and measurable set E, can be expressed as a Young integral in the form

$$\int_E f(x) \, dx = Am(E) + \int_A^B m_E(t, f) \, dt,$$

where

$$A \leqslant f(x) \leqslant B$$

† In Lebesgue's first paper (1902) this adjective is equivalent to 'integrable'.

if $x \in E$, and $m_E(t, f)$ is the measure of the set of points in E at which $f(x) > t$, or in the form

$$\int\limits_E f(x) \, \mathrm{d}x = Bm(E) - \int\limits_A^B \mu_E(t, f) \, \mathrm{d}t,$$

where $\mu_E(t, f)$ is the measure of the set of points in E at which $f(x) < t$.

These expressions cannot be immediately generalized to unbounded functions or to unbounded intervals, for, in the first term either A or B or both will be infinite. This particular difficulty disappears if we restrict ourselves to non-negative functions, for, if $f(x) \geqslant 0$ then

$$\int\limits_E f(x) \, \mathrm{d}x = \int\limits_0^B m_E(t, f) \, \mathrm{d}t.$$

We are now left with the problem of defining the Young integral for an unbounded measurable function $f(x)$ and an unbounded measurable set E.

By Definition 7.5.1 the integrand $m_E(t, f)$ is given in terms of $m_{E_s}(t, f)$ the measure of the set E in the interval $E_s \, (-s \leqslant x \leqslant s)$,

as
$$m_E(t, f) = \lim_{s \to \infty} m_{E_s}(t, f).$$

Hence $m_E(t, f)$ like $m_{E_s}(t, f)$ is a non-decreasing function of t.

The difficulties in framing a definition are that the upper limit B may be infinite and the integrand $m_E(t, f)$ may not be bounded.

However, it is clear that we can restrict ourselves to the case when $m_E(t, f)$ is bounded in any interval

$$0 < \epsilon \leqslant t < \infty.$$

For in the alternative case, for some positive number τ, $m_{E_s}(\tau, f)$ would tend to infinity as $E_s \to \infty$ and so also would $m_{E_s}(t, f)$ for $0 \leqslant t \leqslant \tau$. The Young integral would therefore be infinite. For example, if

$$f(x) = \frac{x}{1+x} \quad \text{for } x \geqslant 0$$

and
$$f(x) = 0 \qquad \text{for } x < 0,$$

then $$0 \leqslant f(x) \leqslant 1,$$

and $$m_{E_s}(t, f) = s - \frac{t}{1-t} \quad \text{if this is positive.}$$

Thus $m_{E_s}(t, f) \to \infty$ as $s \to \infty$ for all $t \geqslant 0$.

If $m_E(t, f)$ is bounded in any interval, $0 < \epsilon \leqslant t$, i.e. if $m_E(t, f)$ is finite for all positive values of t, we can start with the Young integral

$$\int_\epsilon^B m_E(t, f) \, dt$$

and consider the effects of allowing ϵ to tend to zero and B to tend to infinity. We are thus led to the following definition.

DEFINITION 9.2.1. A non-negative function $f(x)$ is 'summable' over the bounded or unbounded, measurable set E, if the Young integral

$$\int_\epsilon^B m_E(t, f) \, dt$$

converges to a finite limit as ϵ tends to zero and B tends independently to infinity.

THEOREM 9.2.1. *If $f(x)$ is non-negative and measurable over a measurable set E, and*
$$f(x) \leqslant \phi(x),$$
where $\phi(x)$ is summable over E, then $f(x)$ is summable over E.

For $$m_E(t, f) \leqslant m_E(t, \phi),$$

and $$\int_\epsilon^B m_E(t, f) \, dt \leqslant \int_\epsilon^B m_E(t, \phi) \, dt.$$

Hence the integral on the left must converge as $\epsilon \to 0$ and $B \to \infty$.

DEFINITION 9.2.2. The Lebesgue integral over E of a non-negative function $f(x)$ summable over E is

$$\int_E f(x) \, dx = \lim_{B \to \infty, \, \epsilon \to 0} \int_\epsilon^B m_E(t, f) \, dt.$$

Consider, for example, the function

$$f(x) = \frac{\sin^2 x}{x^2}, \quad \text{if } x \neq 0, \qquad f(0) = 0,$$

for which $B = 1$. The set of points at which $f(x) > t$ is obviously covered by the set of points at which $x^{-2} > t$. Hence, if E is the semi-infinite interval $0 \leqslant x \leqslant \infty$, then

$$m_E(t, f) < t^{-\frac{1}{2}}$$

and
$$\int_\epsilon^B m_E(t, f) \, \mathrm{d}t < \int_\epsilon^1 t^{-\frac{1}{2}} \, \mathrm{d}t = 2(1 - \epsilon^{\frac{1}{2}}).$$

Thus the function $x^{-2} \sin^2 x$ is summable over $(0 \leqslant x \leqslant \infty)$.

To extend these definitions to functions that take both positive and negative values we introduce the concept of the positive and negative parts of a function.

DEFINITION 9.2.3. The positive and negative parts of a function $f(x)$ defined on a set E are the non-negative functions, defined also on E such that

$$f^+(x) = \begin{cases} f(x) & (f(x) \geqslant 0), \\ 0 & (f(x) < 0), \end{cases}$$

and
$$f^-(x) = \begin{cases} -f(x) & (f(x) < 0), \\ 0 & (f(x) \geqslant 0). \end{cases}$$

THEOREM 9.2.2.

(a) $f(x) = f^+(x) - f^-(x)$, $|f(x)| = f^+(x) + f^-(x)$.

(b) If $f(x)$ is bounded and measurable on a measurable set E, so also are $f^+(x)$ and $f^-(x)$ and

$$\int_E f(x) \, \mathrm{d}x = \int_E f^+(x) \, \mathrm{d}x - \int_E f^-(x) \, \mathrm{d}x,$$

$$\int_E |f(x)| \, \mathrm{d}x = \int_E f^+(x) \, \mathrm{d}x + \int_E f^-(x) \, \mathrm{d}x.$$

Since $f^+(x) = \max(f, 0)$ and $f^-(x) = \max(-f, 0)$, $f^+(x)$ and $f^-(x)$ are bounded and measurable by Theorem 8.2.3, and the expressions for the integrals of $f(x)$ and $|f(x)|$ follow from the addition theorem for functions (8.7.5).

DEFINITION 9.2.4. The function $f(x)$ is summable over E if its positive and negative parts $f^+(x)$ and $f^-(x)$ are each summable over E.

DEFINITION 9.2.5. If $f(x)$ is summable over E its Lebesgue integral over E is

$$\int_E f(x)\,\mathrm{d}x = \int_E f^+(x)\,\mathrm{d}x - \int_E f^-(x)\,\mathrm{d}x.$$

THEOREM 9.2.3. *If $f(x)$ is non-negative and summable over the measurable set I and if I covers the measurable set J, then $f(x)$ is also summable over J, and*

$$\int_I f(x)\,\mathrm{d}x \geqslant \int_J f(x)\,\mathrm{d}x.$$

For the set $E\{x; f(x) > t, x \in J\}$ is the intersection of the measurable sets $E\{x; f(x) > t, x \in I\}$ and J, and hence is measurable. Also

$$E\{x; f(x) > t, x \in I)\} \supset E\{x; f(x) > t, x \in J)\};$$

whence $\qquad\qquad m_I(t, f) \geqslant m_J(t, f),$

and $\qquad\qquad \int\limits_0^\infty m_I(t, f)\,\mathrm{d}t \geqslant \int\limits_0^\infty m_J(t, f)\,\mathrm{d}t.$

Two important consequences of these definitions relate to sets of zero measure.

THEOREM 9.2.4. *If $f(x)$ is any function, bounded or unbounded, measurable or not, and if E is any set of measure zero, then*

$$\int_E f(x)\,\mathrm{d}x = 0.$$

For $\qquad\qquad m_E(t, f^+) \leqslant m(E) = 0,$

and $\qquad\qquad m_E(t, f^-) \leqslant m(E) = 0.$

Hence $\qquad \int\limits_\epsilon^B m_E(t, f^+)\,\mathrm{d}t = 0 = \int\limits_\epsilon^B m_E(t, f^-)\,\mathrm{d}t,$

and therefore $\qquad\qquad \int_E f(x)\,\mathrm{d}x = 0,$

by Definition 9.2.5.

THEOREM 9.2.5. *If $f(x)$ is summable over a measurable set E and Z is a subset of E of measure zero, then*

$$\int_E f(x)\,\mathrm{d}x = \int_F f(x)\,\mathrm{d}x,$$

where F is the complement of Z with respect to E.

We need only prove the theorem for $f^+(x)$.

Now $\qquad\qquad m_E(t, f^+) = m_F(t, f^+) + m_Z(t, f^+)$

and $\qquad\qquad\qquad m_Z(t, f^+) = 0.$

Hence

$$\int_E f^+(x)\,\mathrm{d}x = \int_0^\infty m_E(t, f^+)\,\mathrm{d}t = \int_0^\infty m_F(t, f^+)\,\mathrm{d}t = \int_F f^+(x)\,\mathrm{d}x.$$

Hence in calculating a Lebesgue integral over E we can always neglect the contribution from a set of points in E of measure zero.

DEFINITION 9.2.6. Two functions $f(x)$ and $g(x)$ each summable over E are said to be 'equivalent' in E if $f(x) = g(x)$ almost everywhere in E.

THEOREM 9.2.6. *If $f(x)$ and $g(x)$ are each summable over E and $f(x)$ and $g(x)$ are equivalent in E, then*

$$\int_E f(x)\,\mathrm{d}x = \int_E g(x)\,\mathrm{d}x.$$

9.3. The Lebesgue integral of summable functions as a positive, linear, 'continuous' functional

To justify the definition of the Lebesgue integral of a summable function we must show that, as in the case of a bounded measurable function, it is a positive, linear, continuous functional, but now 'continuity' is taken in the very general sense of the theorem of 'dominated convergence' (9.3.7).

THEOREM 9.3.1. *If $f(x)$ is non-negative and summable over a measurable set E, then*

$$\int_E f(x)\,\mathrm{d}x \geqslant 0.$$

For in this case

$$\int\limits_{E} f(x) \, \mathrm{d}x = \int\limits_{E} f^{+}(x) \, \mathrm{d}x = \int\limits_{0}^{\infty} m_{E}(t, f^{+}) \, \mathrm{d}t \geqslant 0.$$

Hence the Lebesgue integral of a summable function is a positive functional.

THEOREM 9.3.2 (the addition theorem for sets). *If the bounded measurable set E is the union of a finite or enumerable collection of disjoint, measurable sets E_n ($n = 1, 2, \dots$) and if $f(x)$ is summable over each set E_n then it is also summable over E and*

$$\int\limits_{E} f(x) \, \mathrm{d}x = \sum\limits_{n=1}^{\infty} \int\limits_{E_n} f(x) \, \mathrm{d}x.$$

It is clearly sufficient to prove this theorem for $f^{+}(x)$. Let F be the set of points in E at which

$$0 < \epsilon < f(x) < B$$

and F_n be the corresponding set of points in E_n. Then, by Theorems 8.7.2 and 9.2.3,

$$\int\limits_{F} f^{+}(x) \, \mathrm{d}x = \sum\limits_{k=1}^{\infty} \int\limits_{F_k} f^{+}(x) \, \mathrm{d}x \leqslant \sum\limits_{k=1}^{\infty} \int\limits_{E_k} f^{+}(x) \, \mathrm{d}x.$$

Hence, proceeding to the limit $\epsilon \to 0$, $B \to \infty$,

$$\int\limits_{E} f^{+}(x) \, \mathrm{d}x \leqslant \sum\limits_{k=1}^{\infty} \int\limits_{E_k} f^{+}(x) \, \mathrm{d}x.$$

Also
$$\int\limits_{F} f^{+}(x) \, \mathrm{d}x \geqslant \sum\limits_{k=1}^{p} \int\limits_{F_k} f^{+}(x) \, \mathrm{d}x.$$

Hence, proceeding to the same limit,

$$\int\limits_{E} f^{+}(x) \, \mathrm{d}x \geqslant \sum\limits_{k=1}^{p} \int\limits_{E_k} f^{+}(x) \, \mathrm{d}x.$$

Finally, let p tend to infinity. Then

$$\int\limits_{E} f^{+}(x) \, \mathrm{d}x \geqslant \sum\limits_{k=1}^{\infty} \int\limits_{E_k} f^{+}(x) \, \mathrm{d}x.$$

The two bounds we have obtained for the integral of $f^{+}(x)$ over E establish the theorem.

THEOREM 9.3.3 (the addition theorem for summable functions). *If the functions $f(x)$ and $g(x)$ are each summable over the measurable set E, then so also is their sum $\phi(x) = f(x) + g(x)$ and*

$$\int_E \phi(x)\,\mathrm{d}x = \int_E f(x)\,\mathrm{d}x + \int_E g(x)\,\mathrm{d}x.$$

We first prove the theorem for non-negative functions $f(x)$ and $g(x)$.

In the notation of § 8.2 let

$$F = E(0 < \epsilon \leqslant f \leqslant B), \quad G = E(0 < \epsilon \leqslant g \leqslant B).$$

$$I = \text{intersection of } F \text{ and } G, \quad U = \text{union of } F \text{ and } G.$$

Then $I \subseteq F$, $G \subseteq U$.

In the abbreviated notation, $\int_I f = \int_I f(x)\,\mathrm{d}x$, etc., it is clear that

$$\int_I f + \int_I g \leqslant \int_F f + \int_G g \leqslant \int_U f + \int_U g.$$

Now $f(x)$, $g(x)$, and $\phi(x)$ are bounded and measurable in each of the bounded and measurable sets I, F, G, and U. Hence, by Theorem 8.7.5,

$$\int_I f + \int_I g = \int_I \phi,$$

and

$$\int_U f + \int_U g = \int_U \phi.$$

Let $\epsilon \to 0$ and $B \to \infty$. Then

$$\int_F f \to \int_E f \quad \text{and} \quad \int_G g \to \int_E g.$$

Hence $\int_I \phi$ is bounded for all $I \subseteq E$ and therefore ϕ is summable over E. Thus

$$\int_E \phi \leqslant \int_E f + \int_E g \leqslant \int_E \phi;$$

whence the theorem follows for non-negative functions f, g, and ϕ.

In the general cases when $f(x)$ and $g(x)$ may each be of variable sign, we note first that

$$|\phi(x)| \leqslant |f(x)| + |g(x)|.$$

Now $|f(x)|$ and $|g(x)|$ are each summable over E, whence so also is their sum $|f(x)|+|g(x)|$, as we have just proved. Therefore, by Theorem 9.2.1, $|\phi(x)|$ is summable over E, and so also is $\phi(x)$.

We replace the usual argument, due to de la Vallée-Poussin, by the following concise proof, due to John Wright.

By Theorem 9.2.1, ϕ^+ and ϕ^- are each summable over E. Now

$$\phi^++f^-+g^- = \phi^-+f^++g^+;$$

whence, by applying the above result twice,

$$\int_E \phi^+ + \int_E f^- + \int_E g^- = \int_E \phi^- + \int_E f^+ + \int_E g^+.$$

Therefore

$$\int_E \phi = \int_E \phi^+ - \int_E \phi^- = \int_E f + \int_E g.$$

THEOREM 9.3.4. *If $f(x)$ and $g(x)$ are each summable over a measurable set E and if a, b are any real numbers, then*

$$\int_E (af+bg) = a \int_E f + b \int_E g.$$

In the notation of Theorem 9.3.3,

$$\int_E af = \lim \int_F af = a \lim \int_F f = a \int_E f.$$

Hence, by Theorem 9.3.3,

$$a \int_E f + b \int_E g = \int_E af + \int_E bg = \int_E (af+bg).$$

Thus the Lebesgue integral of summable functions is a positive, linear functional.

Clearly there can be no mean value theorem for unbounded functions but there are a number of related theorems.

THEOREM 9.3.5. *If $f(x)$ and $g(x)$ are each summable over a bounded measurable set E, and if $f(x) \geqslant g(x)$ in E, then*

$$\int_E f(x)\,\mathrm{d}x \geqslant \int_E g(x)\,\mathrm{d}x.$$

For, if $f(x)-g(x) = \phi(x)$ then $\phi(x) \geqslant 0$ in E, and, by Theorem 9.3.1,

$$\int_E \phi(x)\,\mathrm{d}x \geqslant 0.$$

Hence, by Theorem 9.3.3,

$$\int_E f(x)\,\mathrm{d}x = \int_E g(x)\,\mathrm{d}x + \int_E \phi(x)\,\mathrm{d}x \geqslant \int_E g(x)\,\mathrm{d}x.$$

THEOREM 9.3.6. *If $f(x)$ is measurable over a measurable set E, and*

$$|f(x)| \leqslant \phi(x),$$

where $\phi(x)$ is summable over E, then $f(x)$ is summable over E.

For if F is the set of points in E at which

$$0 < \epsilon \leqslant f^+(x) \leqslant B,$$

then, by Theorems 9.3.5 and 9.2.3,

$$\int_F f^+(x)\,\mathrm{d}x \leqslant \int_F \phi(x)\,\mathrm{d}x \leqslant \int_E \phi(x)\,\mathrm{d}x.$$

Hence the integral of $f^+(x)$ over F converges as $\epsilon \to 0$, $B \to \infty$. Therefore $f^+(x)$ is summable over E.

Similarly $f^-(x)$ is summable over E, and so also is $f(x)$.

Finally we come to the triumphant climax of this series of theorems—Lebesgue's great theorem of 'dominated convergence'.

THEOREM 9.3.7. *If $\{f_n(x)\}$ $(n = 1, 2, ...)$ is a sequence of measurable functions defined in a bounded measurable set E, if*

$$|f_n(x)| \leqslant \phi(x)$$

for each n, where $\phi(x)$ is summable in E, and if $f_n(x)$ converges pointwise in E as $n \to \infty$ to a function $f(x)$, then $f(x)$ is summable in E and

$$\lim_{n \to \infty} \int_E f_n(x)\,\mathrm{d}x = \int_E f(x)\,\mathrm{d}x.$$

Since $|f_n(x)| \leqslant \phi(x)$ for each n, $|f(x)| \leqslant \phi(x)$, and hence by Theorem 9.3.6, $f(x)$ is summable in E.

As in the proof of Theorem 8.7.7 we follow the concise proof given by de la Vallée-Poussin.

Let $g_n(x) = |f(x) - f_n(x)|$ and let ϵ be any positive number. Then, by Theorems 9.3.4 and 9.2.2, $g_n(x)$ is summable in E. Let

$$E_1 = E\{x; \epsilon > g_1(x), g_2(x),...\}$$

$$E_2 = E\{x; g_1(x) \geqslant \epsilon > g_2(x), g_3(x),...\}$$

$$E_{k+1} = E\{x; g_k(x) \geqslant \epsilon > g_{k+1}(x), g_{k+2}(x)...\}.$$

Then the sets $\{E_k\}$ ($k = 1, 2,...$) form an enumerable collection of disjoint, measurable sets whose union is E. Hence by the addition theorem for summable functions (9.3.3),

$$\int_E g_n = \sum_{k=1}^n \int_{E_k} g_n + \int_{F_n} g_n,$$

where

$$F_n = E - E_1 - E_2 - ... - E_n.$$

Since $0 \leqslant g_n < \epsilon$ in each of the sets $E_1, E_2,..., E_n$ it follows that $0 \leqslant \int_{E_k} g_n \leqslant \epsilon m(E_k)$ if $k = 1, 2,..., n$. Also

$$0 \leqslant \int_{F_n} g_n \leqslant 2 \int_{F_n} \phi,$$

and

$$\int_{F_n} \phi = \int_E \phi - \int_{E_1} \phi - \int_{E_2} \phi - ... - \int_{E_n} \phi,$$

whence

$$\int_{F_n} \phi \to 0 \quad \text{as } n \to \infty.$$

Therefore

$$\limsup_{n \to \infty} \int_E g_n < \epsilon m(E_1) + \epsilon m(E_2) + ... + \epsilon m(E_n) \leqslant \epsilon m(E).$$

Since $m(E)$ is finite and this inequality is true for all $\epsilon > 0$

$$\int_E g_n \to 0 \quad \text{as } n \to \infty.$$

Thus

$$\left| \int_E f_n(x)\, dx - \int_E f(x)\, dx \right| \leqslant \int_E g_n(x)\, dx \to 0 \quad \text{as } n \to \infty.$$

COROLLARY. *The theorem of dominated convergence is also true for any measurable set E, bounded or unbounded.*

Let E_s be the intersection of E and the finite interval $-s \leqslant x \leqslant s$. Then

$$\int_E |f-f_n| = A_{n,s} + B_{n,s},$$

where

$$A_{n,s} = \int_{E-E_s} |f-f_n|$$

and

$$B_{n,s} = \int_{E_s} |f-f_n|.$$

For all n,

$$0 \leqslant A_{n,s} \leqslant 2\int_E \phi - 2\int_{E_s} \phi.$$

Hence there exists an integer $\sigma(\epsilon)$ such that $A_{n,s} \leqslant \epsilon$ for $s \geqslant \sigma(\epsilon)$ and for all n. Also there exists an integer $r(s,\epsilon)$ such that

$$B_{n,s} \leqslant \epsilon \quad \text{for } n \geqslant r(s,\epsilon).$$

Hence there exists an integer $\nu(\sigma,\epsilon)$ dependent only on ϵ such that

$$A_{n,\sigma} + B_{n,\sigma} \leqslant 2\epsilon \quad \text{for } n > \nu(\sigma,\epsilon).$$

Therefore

$$\lim_{n \to \infty} \int_E |f-f_n| = 0$$

and the theorem is established.

Thus the Lebesgue integral of summable functions is a continuous functional.

The existence of a summable dominant function $\phi(x)$ such that $|f_n(x)| \leqslant \phi(x)$ and $\phi(x)$ is summable over an unbounded measured interval E is sufficient but not necessary for the convergence of the sequence, $\int_E f_n(x)\,\mathrm{d}x$ to the limit $\int_E \lim f_n(x)\,\mathrm{d}x$. A simple counter example is given by

$$f_n(x) = \begin{cases} x^{-1} & (n-\tfrac{1}{2} < x < n+\tfrac{1}{2}), \\ 0 & \text{otherwise.} \end{cases}$$

Then

$$f_n(x) \to 0 \quad \text{as } n \to \infty$$

for all x and

$$\int_0^\infty f_n(x)\,\mathrm{d}x = \ln\frac{2n+1}{2n-1} \to 0 \quad \text{as } n \to \infty.$$

But if $\phi(x) \geqslant |f_n(x)|$ for all n, then $\phi(x) \geqslant x^{-1}$ and hence no summable dominant function exists.

9.4. The Lebesgue integral as a primitive

We have already proved in Theorem 8.8.2 that if $f(x)$ is a bounded, measurable function in an interval $[a, b]$ then the indefinite Lebesgue integral

$$\phi(x) = \int_a^x f(t)\, \mathrm{d}t$$

possesses, almost everywhere in $[a, b]$, a derivative $\phi'(x)$ equal to $f(x)$. We can now extend this result to summable functions.

We need a preliminary lemma, due to Fatou, on sequences of functions $\{f_n(x)\}$ which converge to a limit $f(x)$ (but which do *not* possess dominated convergence).

THEOREM 9.4.1 (Fatou's lemma). *If $\{f_n(x)\}$ is any sequence of non-negative functions, each summable in a bounded measurable set E, and if $f_n(x) \to f(x)$ pointwise in E, and $\liminf \int_E f_n(x)\, \mathrm{d}x < \infty$, then $f(x)$ is summable over E and*

$$\liminf_{n \to \infty} \int_E f_n(x)\, \mathrm{d}x \geqslant \int_E f(x)\, \mathrm{d}x.$$

Let

$$f_{n,k} = \min(f_n, k) \quad (k = 1, 2, \ldots).$$

Then, as $n \to \infty$, $f_{n,k} \to \min(f, k) = \phi_k$, say. Hence, by Theorem 8.7.7,

$$\int_E f_{n,k} \to \int_E \phi_k.$$

But $f_{n,k} \leqslant f_n$, whence, by Theorem 9.3.5,

$$\int_E f_{n,k} \leqslant \int_E f_n.$$

Therefore

$$\liminf \int_E f_n \geqslant \liminf \int_E f_{n,k} = \lim \int_E f_{n,k} = \int_E \phi_k.$$

Since this is true for all k and since

$$\int_0^k m(t, f)\, \mathrm{d}t = \int \phi_k \leqslant \liminf \int_E f_n,$$

it follows that f is summable and that

$$\int_E f \leqslant \liminf \int_E f_n.$$

This theorem is the 'best possible' with the prescribed restrictions on the functions $f_n(x)$ for we can easily construct an example where the sign of inequality must be taken.

Let $\qquad f_n(x) = \begin{cases} n^2 x & (0 \leqslant x \leqslant 1/n), \\ 0 & (1/n < x). \end{cases}$

Then $f_n(x) \to 0$ for each value of x. Hence $\int_0^1 f = 0$. But

$$\int_0^1 f_n = \int_0^{1/n} n^2 x \, \mathrm{d}x = \tfrac{1}{2} > \int_0^1 f.$$

A special case of Fatou's lemma (9.4.1) is worthy of note, viz. the 'monotone convergence theorem'.

THEOREM 9.4.2. *If* $\{f_n(x)\}$ *is any monotone, non-decreasing sequence of non-negative functions, each summable in a bounded, measurable set* E, $f_n(x) \to f(x)$ *pointwise in* E, *and*

$$\lim \int_E f_n(x) \, \mathrm{d}x < \infty,$$

then $f(x)$ *is summable over* E *and*

$$\int_E f_n(x) \, \mathrm{d}x \to \int_E f(x) \, \mathrm{d}x.$$

For by Fatou's lemma (9.4.1), $f(x)$ is summable over E. Hence

$$\int_E f_n(x) \, \mathrm{d}x \leqslant \int_E f(x) \, \mathrm{d}x$$

$$\leqslant \liminf_{n \to \infty} \int_E f_n(x) \, \mathrm{d}x \quad \text{(by 9.4.1)}$$

$$= \lim_{n \to \infty} \int_E f_n(x) \, \mathrm{d}x,$$

since the sequence $\left\{ \int_E f(x) \, \mathrm{d}x \right\}$ is monotonic; therefore

$$\lim_{n \to \infty} \int_E f_n(x) \, \mathrm{d}x = \int_E f(x) \, \mathrm{d}x.$$

Before studying the differentiation of an indefinite integral, it is a simpler problem to study the integration of a derivative. The surprising result is given by

THEOREM 9.4.3. *If $\phi(x)$ is continuous and non-decreasing in $[a,b]$ then its derivative $\phi'(x)$ is summable over (a,b) and*

$$\int_a^b \phi'(x)\, dx \leqslant \phi(b)-\phi(a).$$

By Theorem 5.7.5 the incrementary ratio

$$\phi(x, n) = n\phi(x+1/n)-n\phi(x) \quad (n = 1, 2,...)$$

converges to a limit $\phi'(x)$ as $n \to \infty$ almost everywhere in (a, b), i.e. at a set of points E with measure $m(E) = b-a$.

Hence, by Fatou's lemma, $\phi'(x)$ is summable over E and

$$\int_E \phi'(x)\, dx \leqslant \liminf_{n\to\infty} \int_E \phi(x, n)\, dx.$$

Now, by Theorem 9.2.5,

$$\int_E \phi'(x)\, dx = \int_a^b \phi'(x)\, dx,$$

while

$$\int_E \phi(x, n)\, dx = \int_a^b \phi(x, n)\, dx$$

$$= n \int_a^b \phi(x+1/n)\, dx - n \int_a^b \phi(x)\, dx$$

$$= n \int_b^{b+1/n} \phi(x)\, dx - n \int_a^{a+1/n} \phi(x)\, dx$$

if we define $\phi(x)$ as equal to $\phi(b)$ in the interval $b \leqslant x \leqslant b+1/n$. Now

$$n \int_b^{b+1/n} \phi(x)\, dx = \phi(b),$$

and

$$n \int_a^{a+1/n} \phi(x)\, dx \geqslant \phi(a).$$

Hence

$$\int_a^b \phi'(x)\, dx \leqslant \liminf_{n\to\infty} \int_a^b \phi(x, n)\, dx \leqslant \phi(b)-\phi(a).$$

Once again we note that this is the 'best possible' result, for there are continuous and non-decreasing functions $\phi(x)$ for which

$$\int_a^b \phi'(x)\,dx < \phi(b)-\phi(a).$$

To investigate the differentiability of the indefinite Lebesgue integral

$$\phi(x) = \int_a^x f(t)\,dt$$

we can, as usual, restrict ourselves to a non-negative function $f(x)$, but we need to verify the continuity of $\phi(x)$ when $f(x)$ is unbounded but summable.

THEOREM 9.4.4. *If $f(x)$ is non-negative and summable over the interval $[a,b]$, then the Lebesgue integral*

$$\phi(x) = \int_a^x f(t)\,dt$$

is continuous for all x in $[a,b]$.

Let
$$f_n(x) = \begin{cases} f(x) & (f(x) \leqslant n) \\ n & (f(x) > n). \end{cases}$$

Then
$$f_n(x) \to f(x) \quad \text{as } n \to \infty,$$

and, by Theorem 9.3.7,

$$\int_a^b f_n(x)\,dx \to \int_a^b f(x)\,dx \quad \text{as } n \to \infty.$$

Hence to any tolerance $\epsilon > 0$ there corresponds an integer n such that

$$0 \leqslant \int_a^b f(x)\,dx - \int_a^b f_n(x)\,dx < \tfrac{1}{2}\epsilon.$$

Now $f_n(x) \leqslant f(x)$, whence, if $a \leqslant \alpha < \beta \leqslant b$, then

$$0 \leqslant \int_\alpha^\beta f(x)\,dx - \int_\alpha^\beta f_n(x)\,dx < \tfrac{1}{2}\epsilon$$

and
$$0 \leqslant \int_\alpha^\beta f_n(x)\,dx < n(\beta-\alpha)+\tfrac{1}{2}\epsilon.$$

Therefore, if $\beta-\alpha < \epsilon/2n$, then
$$0 < \phi(\beta)-\phi(\alpha) < \epsilon,$$
i.e. $\phi(x)$ is continuous for all x in $[a,b]$.

THEOREM 9.4.5. *If $f(x)$ is summable over the interval $[a,b]$ then the Lebesgue integral $\phi(x) = \int_a^x f(t)\,\mathrm{d}t$ possesses almost everywhere in (a,b) a derivative $\phi'(x)$ equal to $f(x)$.*

The function $f(x)-f_n(x)$ is non-negative, and summable, and the integral
$$\int_a^x \{f(t)-f_n(t)\}\,\mathrm{d}t = \phi(x)-\phi(a)-\int_a^x f_n(t)\,\mathrm{d}t$$
is a non-decreasing, continuous function of x. Hence, by Theorem 5.7.5, it possesses almost everywhere a derivative which is non-negative. By the same theorem the integrals
$$\int_a^x f(t)\,\mathrm{d}t \quad \text{and} \quad \int_a^x f_n(t)\,\mathrm{d}t$$
are differentiable almost everywhere.

Now
$$\phi'(x) \geqslant \frac{\mathrm{d}}{\mathrm{d}x}\int_a^x f_n(t)\,\mathrm{d}t + \overline{\lim_{h\to 0}}\frac{1}{h}\int_a^x \{f(t)-f_n(t)\}\,\mathrm{d}t \quad \text{p.p.}$$
and $f(t) \geqslant f_n(t)$.

Therefore, by Theorem 8.8.2,
$$\phi'(x) \geqslant \frac{\mathrm{d}}{\mathrm{d}x}\int_a^x f_n(t)\,\mathrm{d}t = f_n(x) \quad \text{p.p.}$$
and
$$\phi'(x) \geqslant \lim_{n\to\infty} f_n(x) = f(x) \quad \text{p.p.}$$

But, by Theorem 9.3.3,
$$\int_a^b \{\phi'(x)-f(x)\}\,\mathrm{d}x = \int_a^b \phi'(x)\,\mathrm{d}x - \int_a^b f(x)\,\mathrm{d}x \leqslant 0.$$

Hence, by the null integral theorem (8.7.4),
$$\phi'(x) = f(x) \quad \text{p.p. in } (a,b).$$

9.5. Exercises

1. Show that the function

$$f(x) = \begin{cases} \dfrac{\sin x}{x} & (x \neq 0) \\ 1 & (x = 0) \end{cases}$$

is not summable over the interval $(0 \leqslant x \leqslant \infty)$.

2. If the non-negative functions $f_n(x)$ $(n = 1, 2, ...)$ are each summable over a measurable set E, and if $f_n(x) \leqslant f_{n+1}(x)$ prove that the limit function

$$f(x) = \lim_{n \to \infty} f_n(x)$$

is summable over E and that

$$\int_E f_n(x) \, dx \to \int_E f(x) \, dx \quad \text{as } n \to \infty.$$

3. If $f_n(x)$ denotes the truncated function

$$f_n(x) = \begin{cases} f(x) & \text{if } 0 \leqslant x \leqslant n \\ n & \text{if } n < x \end{cases} \quad (n = 1, 2, ...)$$

prove that $(f+g)_n \leqslant f_n + g_n \leqslant (f+g)_{2n}$, and deduce the addition theorem (9.3.3) for functions.

4. The function

$$f(x) = 2x \sin(1/x^2) - (2/x)\cos(1/x^2)$$

is the derivative of $x^2 \sin(1/x^2)$. Explain why $f(x)$ is not summable over $[0, 1]$.

5. Examine the sequence of integrals $\int\limits_0^\infty f_n(x) \, dx$ in the light of the theorems of dominated and monotone convergence if

  (i) $f_n(x) = nxe^{-nx^2}$,
  (ii) $f_n(x) = 2n^2 e^{-n^2 x^2}$,
  (iii) $f_n(x) = \phi(nx)$,
  (iv) $f_n(x) = \phi(x-n)$,

where $\phi(x)$ is summable over $(0, \infty)$.

10 Multiple integrals

10.1. Introduction

It is possible to develop the theory of Lebesgue measure and integration in d-dimensional Euclidean space from the very beginning, but in the interests of intelligibility we have so far restricted our exposition to one dimension. We have, however, so phrased the terminology, the notation, the definitions, and the theorems so that most of them are applicable to d-dimensional space.

10.2. Elementary sets in d dimensions

DEFINITION 10.2.1. In d dimensions, with coordinates $(x_1, x_2,..., x_d)$ an 'interval' is the Cartesian product of the linear intervals
$$a_k < x_k < b_k \quad (k = 1, 2,..., d)$$
where a_k, b_k are finite or infinite numbers for all k.

DEFINITION 10.2.2. An 'elementary set', with indicator σ is the union of a finite number of disjoint intervals, with indicators $\sigma_1, \sigma_2,..., \sigma_n$ so that
$$\sigma_j \sigma_k = 0 \quad \text{if } j \neq k,$$
and
$$\sigma = \sigma_1 + \sigma_2 + ... + \sigma_n.$$
The intervals $\{\sigma_s\}$ are called the 'components 'of σ.

THEOREM 10.2.1. *The intersection, union, difference, and symmetric difference of two elementary sets are also elementary sets.*

The intersection of two intervals
$$(a_k < x_k < b_k) \quad \text{and} \quad (p_k < x_k < q_k) \quad (k = 1, 2,..., d)$$

is an interval of the form

$$\max(a_k, p_k) \prec x_k \prec \min(b_k, q_k) \quad (k = 1, 2, ..., d).$$

The proof for the intersection of two elementary sets then follows as in Theorem 6.2.1.

If the interval τ is covered by an interval ω, then the complement $\omega - \tau$ is an elementary set. For if τ is the interval $(a_k \prec x_k \prec b_k)$ $(k = 1, 2, ..., d)$ then the $2d$ hyperplanes, $x = a_k$ and $x = b_k$ divide the interval ω into $3d$ intervals, one of which may be taken to be τ and the remainder of which form an elementary set, i.e. the complement $\omega - \tau$ is an elementary set.

If the elementary set τ is covered by another elementary set σ then the complement $\sigma - \tau$ is an elementary set. For

$$\sigma - \tau = \sigma - \sigma\tau$$
$$= \sum_s \sigma_s(1 - \tau),$$

where $\{\sigma_s\}$ are the component intervals of σ. Now σ_s and $1 - \tau$ are elementary sets, and so is their intersection $\sigma_s(1 - \tau)$. Also $\sigma_s(1 - \tau) \cdot \sigma_t(1 - \tau) = 0$, i.e. the elementary sets $\sigma_\sigma(1 - \tau)$ are disjoint. Thus $\sigma - \tau$ is the union of a finite number of disjoint intervals and is therefore an elementary set.

It follows as in Theorem 6.2.1 that, if σ and τ are elementary sets, so also is their union $\sigma \cup \tau$, the differences $\sigma - \sigma\tau$, $\tau - \sigma\tau$, and the symmetric difference $\sigma \triangle \tau$. Thus elementary sets in d dimensions form an algebra, as in one dimension.

THEOREM 10.2.2. *If σ and τ are any pair of elementary sets then there exists a finite collection of disjoint intervals $\{\gamma_s\}$ such that*

$$\sigma = \sum_s a_s \gamma_s,$$
$$\tau = \sum_t b_t \gamma_t,$$
$$\sigma \cup \tau = \sum_s \gamma_s,$$

where each coefficient a_s or b_t is either 0 or 1.

The proof is exactly the same as in Theorem 6.2.2.

DEFINITION 10.2.3. The geometric measure $g(\alpha)$ of an interval α $(a_k \prec x_k \prec b_k)$ $(k = 1, 2, ..., d)$ is its area (if $d = 2$), volume (if $d = 3$), or hypervolume (if $d > 3$), and

$$g(\alpha) = \prod_{k=1}^{d} (b_k - a_k).$$

THEOREM 10.2.3. *If the interval ω is the union of a finite number N of disjoint intervals $\{\sigma_s\}$ then*

$$g(\omega) = \sum_{s=1}^{N} g(\sigma_s).$$

If σ_s is the interval $(a_k^{(s)} \prec x_k \prec b_k^{(s)})$ $(k = 1, 2, ..., d)$ then the $2dN$ hyperplanes, $x_k = a_k^{(s)}$, $x_k = b_k^{(s)}$ divide the interval ω into a finite number of intervals, $\omega_1, \omega_2, ...$ which can be enumerated so that the interior of ω_s coincides with the interior of σ_s for $s = 1, 2, ..., N$. Hence

$$g(\omega) = \sum_{s=1}^{N} g(\omega_s) = \sum_{s=1}^{N} g(\sigma_s).$$

10.3. Lebesgue theory in *d* dimensions

From this point onwards the d-dimensional theory follows, almost word for word, the one-dimensional theory of §§ 6.2, 6.3, 6.4, 6.5, and 6.6. Outer and inner measure are then defined and discussed exactly as in Chapter 7, and the Lebesgue integral as in Chapters 8 and 9, with one exception noted below and with the convention that the symbols

$$\iint \cdots \int_R f(x) \, dx_1 \, dx_2 \ldots dx_d, \quad \int_a^b f(x) \, dx, \quad \text{or} \quad \int_R f(x)$$

now represent the integral of $f(x_1, x_2, ..., x_d)$ over the interval R

$$a_k \leqslant x_n \leqslant b_k \quad (k = 1, 2, ..., d).$$

The reduction of the d-dimensional theory to the one-dimensional case is facilitated by Lebesgue's concept of integration over a measurable set and by Young's integral

$$\int_E f(x) \, dx = \int m_E(t, f) \, dt,$$

which reduces the d-dimensional integral of $f(x)$ over E to the one-dimensional integral of its measure function over the range of $f(x)$.

The one exception is that we must exclude the results concerning the differentiability of the indefinite Lebesgue integral $\phi(t) = \int_a^t f(x) \, dx$, which is crucially dependent on the monotone and continuous character of $\phi(t)$ as a function of the single variable t for non-negative functions $f(x)$.

There is, however, one new problem that arises in the multi-dimensional theory, which is most clearly exhibited in the case of two dimensions. This is the problem of the relation between the integral

$$\int_R f(x,y) \, dxdy$$

over the rectangle R ($a \leqslant x \leqslant b$, $p \leqslant y \leqslant q$) and the integrals

$$g(y) = \int_a^b f(x,y) \, dx, \qquad h(x) = \int_p^q f(x,y) \, dy,$$

$$\int_p^q g(y) \, dy, \qquad \int_a^b h(x) \, dx.$$

Elementary calculus suggests that if $f(x,y)$ is bounded and continuous in R, then the multiple integral can be expressed as a repeated integral in the forms

$$\int_R f(x,y) \, dxdy = \int_p^q g(y) \, dy = \int_a^b h(x) \, dx,$$

but there are a number of well-known examples which show that these relations are not true for all unbounded functions:

(1)   If $\qquad f(x,y) = (x^2-y^2)/(x^2+y^2)^2$

except at the origin, where $f = 0$ and if R is the square

$$(0 \leqslant x \leqslant 1, \quad 0 \leqslant y \leqslant 1),$$

then $\qquad g(y) = -(1+y^2)^{-1}, \qquad h(x) = (1+x^2)^{-1},$

$$\int_0^1 g(y) \, dy = -\tfrac{1}{4}\pi, \qquad \int_0^1 h(x) = \tfrac{1}{4}\pi.$$

The set of points (x, y) at which $f(x, y) > t > 0$, occupies the interior of half one of the loops of the lemniscate,

$$tr^2 = \cos 2\theta,$$

whence
$$m(t, f) = \tfrac{1}{4} t^{-1}$$

and the Young integral,

$$\int\limits_0^\infty m(t, f)\, \mathrm{d}t$$

is not convergent, so that $f(x, y)$ is not summable over R.

(2) If $\phi(z) = pz^p(1 + z^{2p})$ $(p > 1)$, $z = xy$, $f(x, y) = \mathrm{d}\phi(z)/\mathrm{d}z$ and R is the rectangle $(0 \leqslant x \leqslant \infty, 0 \leqslant y \leqslant c)$ then

$$g(y) = 0, \qquad h(x) = x^{-1}\phi(cx),$$

$$\int\limits_0^c g(y)\, \mathrm{d}y = 0. \qquad \int\limits_0^\infty h(x)\, \mathrm{d}x = \tfrac{1}{4}\pi.$$

(Hobson, vol. ii, p. 351).

If $f^+(x, y)$ and $f^-(x, y)$ are the positive and negative parts of $\phi'(z)$ then

$$\int\limits_0^\infty f^+\, \mathrm{d}x = \tfrac{1}{2} p/y = \int\limits_0^\infty f^-\, \mathrm{d}x \quad (\text{if } y > 0).$$

Hence the integral of $\phi'(z)$ is not absolutely convergent.

Similar problems arise in the case of integrals over three or more dimensions but for the sake of clarity we shall restrict our exposition to integrals in two dimensions.

10.4. Fubini's theorem stated

We begin by giving a general statement of the theory which is due to Fubini.

We consider a function $f(x, y)$ which is non-negative and summable over the whole of the (x, y)-plane. To apply the theorem to a summable function taking both positive and negative signs we have only to consider its positive and negative parts separately. Again to apply the theorem to the integral of

$f(x, y)$ over a bounded measurable set E we introduce the function $\phi(x, y)$ defined by the relations

$$\phi(x, y) = \begin{cases} f(x, y) & (x, y \in E), \\ 0 & (x, y \notin E). \end{cases}$$

With the non-negative, summable function $f(x, y)$ we associate its indicator function

$$\alpha(x, y, \sigma) = \begin{cases} 1 & (f(x, y) > \sigma), \\ 0 & (f(x, y) \leqslant \sigma), \end{cases}$$

and its two-dimensional measure function

$$m(\sigma) = \iint \alpha(x, y, \sigma) \, dx dy.$$

We prove that if y is a fixed number then $f(x, y)$, regarded as a function of x, is summable for almost all values of y and has a one-dimensional measure function

$$\eta(y, \sigma) = \int \alpha(x, y, \sigma) \, dx.$$

Similarly we prove that if x is a fixed number then $f(x, y)$, regarded as a function of y, is summable for almost all values of x and has a one-dimensional measure function

$$\xi(x, \sigma) = \int \alpha(x, y, \sigma) \, dy.$$

We then prove that

$$\int \xi(x, \sigma) \, dx = m(\sigma) = \int \eta(y, \sigma) \, dy.$$

Turning to the function $f(x, y)$ we next prove that the integrals

$$g(y) = \int f(x, y) \, dx$$

and

$$h(x) = \int f(x, y) \, dy$$

exist for almost all values of y and x respectively, and that

$$\int g(y) \, dy = \int f(x, y) \, dx dy = \int h(x) \, dx.$$

Fubini's theorem presupposes the existence of the double integral $\iint f(x, y) \, dx dy$ and then expresses this double integral as a pair of repeated integrals

$$I = \int dy \left\{ \int f(x, y) \, dx \right\} \quad \text{and} \quad J = \int dx \left\{ \int f(x, y) \, dy \right\}.$$

But, in many problems of analysis, we can presuppose the existence of one repeated integral I and we require conditions under which I is equal to the other repeated integral J. Sufficient conditions are provided by Tonelli's theorem in the form that if $f(x,y)$ is non-negative and measurable over the (x,y)-plane, and if the integral I exists, then so also does the integral J and $I = J$.

10.5. Fubini's theorem for indicators

THEOREM 10.5.1. *If the set of points E with indicator $\alpha(x,y)$ has finite two-dimensional measure*

$$\mu = \iint \alpha(x,y)\, dxdy$$

then the set of points A on the line $x = s$ with indicator $\alpha(s,y)$ has finite one-dimensional measure

$$\mu(s) = \int \alpha(s,y)\, dy$$

for almost all values of s. Also $\mu(s)$ is integrable over s and

$$\int \mu(s)\, ds = \mu.$$

(i) If E is the interval $(a \prec x \prec b,\ p \prec y \prec q)$, then

$$\mu(s) = \begin{cases} (q-p) & \text{if } a \prec s \prec b, \\ 0 & \text{otherwise} \end{cases}$$

and
$$\int \mu(s)\, ds = (b-a)(q-p) = \mu.$$

(ii) If E is an elementary set $\alpha(x,y,E)$, i.e. the union of a finite collection of disjoint intervals $\{E_n\}$, then E_n will have a one-dimensional measure

$$\mu_n(s) = \int \alpha(s,y,E_n)\, dy$$

and a two-dimensional measure

$$\mu_n = \int \mu_n(s)\, ds.$$

E has the one-dimensional measure

$$\mu(s) = \int \alpha(s,y,E)\, dy = \sum \mu_n(s)$$

and the two-dimensional measure

$$\mu = \iint \alpha(s,y,E)\, dsdy = \iint \sum \alpha(s,y,E_n)\, dsdy.$$

Hence
$$\mu = \int \mu(s)\,\mathrm{d}s.$$

(iii) If E is a bounded, outer set, i.e. the union of an enumerable collection of disjoint intervals $\{E_n\}$ ($n = 1, 2,...$), then each E_n has a one-dimensional measure $\mu(s, E_n)$ and, by Theorem 7.4.3, in two dimensions, E also has a one-dimensional measure

$$\mu(s, E) = \sum_{n=1}^{\infty} \mu(s, E_n).$$

Hence, by the theorem of bounded convergence,

$$\int \mu(s, E)\,\mathrm{d}s = \sum_{n=1}^{\infty} \int \mu(s, E_n)\,\mathrm{d}s = \sum_{n=1}^{\infty} \mu(E_n), \quad \text{by (ii)},$$
$$= \mu(E).$$

(iv) If F is an inner set, the complement of an outer set with respect to a bounded interval I ($a \prec x \prec b,\ p \prec y \prec q$) then

$$\mu(s, F) = \int \{\alpha(s, y, I) - \alpha(s, y, E)\}\,\mathrm{d}y$$
$$= \begin{cases} (q - p) - \mu(s, E) & \text{if } s \in I, \\ 0 & \text{otherwise.} \end{cases}$$

Hence
$$\int \mu(s, F)\,\mathrm{d}s = \int_I (q - p)\,\mathrm{d}s - \int_I \mu(s, E)\,\mathrm{d}s$$
$$= \mu(I) - \mu(E)$$
$$= \mu(F).$$

(v) If G is a bounded, measurable set, then there exist two monotone sequences of bracketing sets $\{E_n\}$ and $\{F_n\}$ such that each E_n is an outer set, each F_n is an inner set, and

$$E_n \supset E_{n+1} \supset G \supset F_{n+1} \supset F_n,$$
and
$$\mu(E_n) - \mu(F_n) < \epsilon_n,$$

where $\{\epsilon_n\}$ is any sequence of positive numbers converging to zero as $n \to \infty$.

Let $\mu(s, E_n)$ and $\mu(s, F_n)$ be one-dimensional measures of E_n and F_n respectively, and let

$$\mu_n(s) = \mu(s, E_n) - \mu(s, F_n).$$

Then
$$\mu_n(s) \geqslant \mu_{n+1}(s) \geqslant 0.$$

Hence, as $n \to \infty$, $\mu_n(s)$ converges to a non-negative function $\mu(s)$. Now

$$\int \mu_n(s) \, \mathrm{d}s = \int \mu_n(s, E_n) \, \mathrm{d}s - \int \mu_n(s, F_n) \, \mathrm{d}s$$
$$= \mu(E_n) - \mu(F_n), \quad \text{by (iv)},$$
$$< \epsilon.$$

Therefore, by the monotone convergence theorem,

$$\int \mu(s) \, \mathrm{d}s = \lim_{n \to \infty} \int \mu_n(s) \, \mathrm{d}s = 0.$$

But $\mu(s) \geqslant 0$. Hence, by the null integral theorem (8.7.4),

$$\mu(s) = 0 \quad \text{p.p.}$$

Now $\qquad \alpha(s, y, F_n) \leqslant \alpha(s, y, G) \leqslant \alpha(s, y, E_n)$
and

$$\int \alpha(s, y, E_n) \, \mathrm{d}y - \int \alpha(s, y, F_n) \, \mathrm{d}y = \mu(s, E_n) - \mu(s, F_n) = \mu_n(s).$$

Therefore for almost all s, $\alpha(s, y, G)$ is integrable with respect to y, i.e. for almost all s, G has a one-dimensional measure $\mu(s, G)$ and

$$\mu(s, F_n) \leqslant \mu(s, G) \leqslant \mu(s, E_n).$$

Hence, $m(s, G)$ is integrable with respect to s and

$$\mu(F_n) = \int \mu(s, F_n) \, \mathrm{d}s \leqslant \int \mu(s, G) \, \mathrm{d}s \leqslant \int \mu(s, E_n) \, \mathrm{d}s = \mu(E_n).$$

Therefore $\qquad \int \mu(s, G) \, \mathrm{d}s = \mu(G).$

(vi) If H is an unbounded, measurable set, let G_n be the intersection of H with the interval $-n \leqslant x \leqslant n$, $-n \leqslant y \leqslant n$. Then

$$\mu(H) = \lim_{n \to \infty} \mu(G_n) = \lim_{n \to \infty} \int \mu(s, G_n) \, \mathrm{d}s$$
$$= \int \lim_{n \to \infty} \mu(s, G_n) \, \mathrm{d}s = \int \mu(s, H) \, \mathrm{d}s.$$

This completes the proof of Fubini's theorem for indicators of sets E with finite two-dimensional measure.

10.6. Fubini's theorem for summable functions

THEOREM 10.6.1. *If $f(x, y)$ is non-negative, bounded, and measurable over the interval I $(a \leqslant x \leqslant b, p \leqslant y \leqslant q)$, then*

$f(x, y)$ is also measurable over the line ($a \leqslant x \leqslant b$, $y =$ constant)
for almost all values of y, and if

$$g(y) = \int_a^b f(x, y) \, dx \quad \text{(p.p. in } y)$$

then $g(y)$ is measurable over the line $p \leqslant y \leqslant q$ and

$$\int_p^q g(y) \, dy = \iint_I f(x, y) \, dx dy.$$

Fubini's theorem for indicators (10.5.1) applies to the indicator $\alpha(x, y, t)$ of the function $f(x, y)$ which has the integral

$$m(t) = \iint_I \alpha(x, y, t) \, dx dy,$$

which is the two-dimensional measure of the set

$$E\{x, y; f(x, y) > t\}.$$

Hence the integral

$$\eta(y, t) = \int_a^b \alpha(x, y, t) \, dx$$

exists for almost all y and gives the measure of the set

$$E\{x; f(x, y) > t, y = \text{constant}\}.$$

Also
$$m(t) = \int_p^q \eta(y, t) \, dy.$$

Now consider the Lebesgue bracketing function $\lambda(x, y)$ of $f(x, y)$ defined in 8.5.1 in the form

$$\lambda(x, y) = \sum_{k=0}^{n-1} t_k \{\alpha(x, y, t_k) - \alpha(x, y, t_{k+1})\}.$$

Since Fubini's theorem applies to the indicator $\alpha(x, y, t_k)$ it follows that the integral

$$\int_a^b \lambda(x, y) \, dx$$

exists for almost all y and that

$$\int_p^q dy \int_a^b \lambda(x, y) \, dx = \iint_I \lambda(x, y) \, dx dy.$$

Similarly if $\mu(x,y)$ is the upper bracketing function

$$\int_p^q \mathrm{d}y \int_a^b \mu(x,y)\,\mathrm{d}x = \iint_I \mu(x,y)\,\mathrm{d}x\mathrm{d}y.$$

Now
$$\lambda(x,y) \leqslant f(x,y) \leqslant \mu(x,y)$$

and

$$\int_a^b \mu(x,y)\,\mathrm{d}x - \int_a^b \lambda(x,y)\,\mathrm{d}x < \epsilon(b-a), \quad \text{p.p. in } y,$$

where
$$\epsilon = \max(t_{k+1}-t_k) \quad (k = 0,1,...,n-1).$$

Hence $\int_a^b f(x,y)\,\mathrm{d}x$ exists for almost all values of y and

$$\int_a^b \lambda(x,y)\,\mathrm{d}x \leqslant \int_a^b f(x,y)\,\mathrm{d}x \leqslant \int_a^b \mu(x,y)\,\mathrm{d}x.$$

Also

$$\int_p^q \mathrm{d}y \int_a^b \lambda(x,y)\,\mathrm{d}x \leqslant \int_p^q \mathrm{d}y \int_a^b f(x,y)\,\mathrm{d}x \leqslant \int_p^q \mathrm{d}y \int_a^b \mu(x,y)\,\mathrm{d}x.$$

Therefore
$$\int_p^q \mathrm{d}y \int_a^b f(x,y)\,\mathrm{d}x = \iint_F f(x,y)\,\mathrm{d}x\mathrm{d}y.$$

THEOREM 10.6.2. *If $f(x,y)$ is non-negative and summable over the (x,y)-plane, I, then $f(x,y)$ is also summable over the line $(-\infty < x < \infty, y = \text{constant})$ for almost all values of y, and if*

$$g(y) = \int_{-\infty}^{\infty} f(x,y)\,\mathrm{d}x$$

then $g(y)$ is summable over the line $-\infty < y < \infty$ and

$$\int_{-\infty}^{\infty} g(y)\,\mathrm{d}y = \iint f(x,y)\,\mathrm{d}x\mathrm{d}y.$$

Define the truncated function $f_n(x,y)$ as follows:

$$f_n = f, \quad \text{if } |x| \leqslant n, \quad |y| \leqslant n, \quad \text{and} \quad 0 \leqslant f \leqslant n,$$
$$f_n = n, \quad \text{if } |x| \leqslant n, \quad |y| \leqslant n, \quad \text{and} \quad f \geqslant n,$$
$$f_n = 0, \quad \text{if } |x| > n \quad \text{or} \quad |y| > n.$$

Then $f_n(x, y)$ is bounded and measurable and, by Theorem 10.6.1, the integral

$$g_n(y) = \int_{-\infty}^{\infty} f_n(x, y) \, \mathrm{d}x$$

exists for almost all y and

$$\int_{-\infty}^{\infty} g_n(y) \, \mathrm{d}y = \iint f_n(x, y) \, \mathrm{d}x\mathrm{d}y.$$

Now, by the theorem of monotone convergence (9.4.2), the integral

$$g(y) = \int_{-\infty}^{\infty} f(x, y) \, \mathrm{d}x$$

exists as the limit

$$\int_{-\infty}^{\infty} \lim_{n\to\infty} f_n(x, y) \, \mathrm{d}x = \lim_{n\to\infty} \int_{-\infty}^{\infty} f_n(x, y) \, \mathrm{d}x = \lim_{n\to\infty} g_n(y).$$

Similarly the integral

$$\iint_I f(x, y) \, \mathrm{d}x\mathrm{d}y$$

exists as the limit

$$\iint_I \lim_{n\to\infty} f_n(x, y) \, \mathrm{d}x\mathrm{d}y = \lim_{n\to\infty} \iint_I f_n(x, y) \, \mathrm{d}x\mathrm{d}y$$

$$= \lim_{n\to\infty} \int_{-\infty}^{\infty} g_n(y) \, \mathrm{d}y$$

$$= \int_{-\infty}^{\infty} \lim_{n\to\infty} g_n(y) \, \mathrm{d}y$$

$$= \int_{-\infty}^{\infty} g(y) \, \mathrm{d}y.$$

This completes the proof of Fubini's theorem.

10.7. Tonelli's theorem

Fubini's theorem can be used to establish the following result, due to Tonelli, which is often of much more practical utility.

THEOREM 10.7.1. *If $f(x, y)$ is a non-negative, measurable function of (x, y) over the x, y plane I ($-\infty < x < \infty$, $-\infty < y < \infty$) and, if one repeated integral*

$$\int\limits_{-\infty}^{\infty} \mathrm{d}y \int\limits_{-\infty}^{\infty} f(x, y) \, \mathrm{d}x$$

exists, then so does the other repeated integral

$$\int\limits_{-\infty}^{\infty} \mathrm{d}x \int\limits_{-\infty}^{\infty} f(x, y) \, \mathrm{d}y$$

and the two repeated integrals are equal.

If $f_n(x, y)$ is the truncated function of Theorem 10.6.2 then $f_n(x, y)$ is non-negative, bounded, and measurable, and the integrals

$$\iint\limits_{I} f_n \, \mathrm{d}x\mathrm{d}y, \qquad \int\limits_{-\infty}^{\infty} \mathrm{d}y \int\limits_{-\infty}^{\infty} f_n \, \mathrm{d}x, \qquad \int\limits_{-\infty}^{\infty} \mathrm{d}x \int\limits_{-\infty}^{\infty} f_n \, \mathrm{d}y$$

each exist and are equal.

The existence of the repeated integral

$$\int\limits_{-\infty}^{\infty} \mathrm{d}y \int\limits_{-\infty}^{\infty} f \, \mathrm{d}x$$

implies the existence of

$$g(y) = \int\limits_{-\infty}^{\infty} f(x, y) \, \mathrm{d}x$$

for almost all y, and the existence of $\int\limits_{-\infty}^{\infty} g(y) \, \mathrm{d}y$.

Now, by the monotone convergence theorem (9.4.2),

$$\int\limits_{-\infty}^{\infty} g(y) \, \mathrm{d}y = \int\limits_{-\infty}^{\infty} \lim_{n \to \infty} g_n(y) \, \mathrm{d}y = \lim_{n \to \infty} \int\limits_{-\infty}^{\infty} g_n(y) \, \mathrm{d}y$$

$$= \lim_{n \to \infty} \iint\limits_{I} f_n(x, y) \, \mathrm{d}x\mathrm{d}y = \iint\limits_{I} \lim_{n \to \infty} f_n(x, y) \, \mathrm{d}x\mathrm{d}y$$

$$\text{by (10.6.1),}$$

i.e. the integral $\iint\limits_{I} f \, \mathrm{d}x\mathrm{d}y$ exists and equals $\int\limits_{-\infty}^{\infty} g(y) \, \mathrm{d}y$.

We can now use Theorem 10.6.2 to establish the existence of

$$h(x) = \int\limits_{-\infty}^{\infty} f(x, y) \, \mathrm{d}y$$

almost everywhere in x and to prove that

$$\iint_I f(x,y)\ \mathrm{d}x\mathrm{d}y = \int_{-\infty}^{\infty} h(x)\ \mathrm{d}x.$$

Combining these results we see that $\int_{-\infty}^{\infty} h(x)\ \mathrm{d}x = \int_{-\infty}^{\infty} g(y)\ \mathrm{d}y$, which is the theorem to be proved.

10.8. Product sets

In order to discuss the geometric definition of the Lebesgue integral (§ 10.9) we shall need to consider the measurability of the 'product set', $G = E \times F$, of a set E on the x-axis and a set F on the y-axis.

DEFINITION 10.8.1. The product set $G = E \times F$, is the set of points with plane Cartesian coordinates (x, y) such that $x \in E$ and $y \in F$.

DEFINITION 10.8.2. If $\alpha \equiv \alpha(x)$ and $\beta \equiv \beta(y)$ are the indicators of E and F respectively, then the indicator of G will be denoted by $\alpha * \beta$.

(The symbol $\sigma\tau$ has been used (Definition 4.2.1) to denote the intersection of the two *collinear* sets σ and τ. To prevent any misunderstanding we shall use the symbol $\alpha * \beta$ to denote the 'Cartesian product' of the set $\alpha(x)$ on the x-axis and the set $\beta(y)$ on the y-axis.)

THEOREM 10.8.1. *If the sets E and F are bounded and measurable, so also is $G = E \times F$, and*

$$m(G) = m(E)m(F).$$

Let $\alpha = \alpha(x)$ and $\beta = \beta(y)$ be the indicators of E and F respectively. Then to any tolerance $\epsilon > 0$ there correspond outer sets $\sigma = \sigma(x)$ and $\tau = \tau(y)$ such that $\alpha \leqslant \sigma$, $\beta \leqslant \tau$ and

$$m(\sigma) - \epsilon \leqslant m(\alpha) \leqslant m(\sigma),$$
$$m(\tau) - \epsilon \leqslant m(\beta) \leqslant m(\tau).$$

Now σ and τ are respectively the unions of enumerable disjoint intervals $\{\sigma_p\}$, $\{\tau_q\}$, so that

$$\sigma = \sum \sigma_p, \quad \tau = \sum \tau_q,$$

and $$\sigma * \tau \leqslant \sum \sigma_p * \tau_q.$$

$\sigma_p * \tau_q$ is a rectangle whose edges are the intervals σ_p and τ_q. Hence

$$m(\sigma_p * \tau_q) = m(\sigma_p)m(\tau_q)$$

and

$$\begin{aligned} m(\sigma * \tau) &= \sum m(\sigma_p)m(\tau_q) \\ &= \sum m(\sigma_p) \sum m(\tau_q) \\ &= m(\sigma)m(\tau). \end{aligned}$$

The product set $\alpha * \beta$ is covered by the open set $\sigma * \tau$, whence

$$m^*(\alpha * \beta) \leqslant m(\sigma * \tau) = m(\sigma)m(\tau) \leqslant \{m(\alpha)+\epsilon\}\{m(\beta)+\epsilon\}.$$

This is true for each $\epsilon > 0$, therefore

$$m^*(\alpha * \beta) \leqslant m(\alpha)m(\beta).$$

Now the sets α and β are each bounded. Hence there are intervals A and B, lying on the x- and y-axes respectively, such that $\alpha \leqslant \sigma \leqslant A$ and $\beta \leqslant \tau \leqslant B$.

The complement of the set $\alpha * \beta$ with respect to the rectangle $A * B$ is expressible as the union of three disjoint sets as

$$A * B - \alpha * \beta = (A-\alpha) * (B-\beta) + (A-\alpha) * \beta + \alpha * (B-\beta).$$

By the result established above for outer measures

$$\begin{aligned} m^*\{(A-\alpha) * (B-\beta)\} &\leqslant m(A-\alpha)m(B-\beta), \\ m^*\{(A-\alpha) * \beta\} &\leqslant m(A-\alpha)m(\beta), \\ m^*\{\alpha * (B-\beta)\} &\leqslant m(\alpha)m(B-\beta). \end{aligned}$$

Therefore

$$m^*(A * B - \alpha * \beta) \leqslant m(A)m(B) - m(\alpha)m(\beta),$$

and

$$\begin{aligned} m_*(\alpha * \beta) &= m(A * B) - m^*(A * B - \alpha * \beta) \\ &\geqslant m(\alpha)m(\beta). \end{aligned}$$

Thus

$$m(\alpha)m(\beta) \leqslant m_*(\alpha * \beta) \leqslant m^*(\alpha * \beta) \leqslant m(\alpha)m(\beta).$$

Hence $\alpha * \beta$ is measurable and $m(\alpha * \beta) = m(\alpha)m(\beta)$.

10.9. The geometric definition of the Lebesgue integral

The definition of the Lebesgue integral given by its inventor in his thesis and first paper (1902) was geometric rather than analytic in character, and it provides an illuminating approach, as can be seen from the clear and concise account given by Burkill (1953). The relation between the geometric and analytic definitions is given by the following definition and theorem.

DEFINITION 10.9.1. If $f(x)$ is a non-negative function defined in a set E, then the 'ordinate set' of $f(x)$ over E is the set of points (x, y) such that $x \in E$ and $0 \leqslant y < f(x)$.

THEOREM 10.9.1. *If $f(x)$ is non-negative, bounded, and measurable over a bounded, measurable set E, then the Lebesgue integral $\int_E f(x) \, dx$ is equal to the two-dimensional measure of the ordinate set of $f(x)$ over E.*

In the notation of § 8.5 let the range $[0, B]$ of $f(x)$ be divided by the finite number of points

$$0 = t_0 < t_1 < t_2 < \ldots < t_n = B.$$

Let $\alpha(x, y)$ be the indicator of the ordinate set of $f(x)$ over E, and let $\eta_p(y)$ be the indicator of the set $t_p \leqslant y < t_{p+1}$.

Let
$$\lambda(x, y) = \sum_{p=0}^{n-1} \alpha(x, t_{p+1}) * \eta_p(y)$$

and
$$\mu(x, y) = \sum_{p=0}^{n-1} \alpha(x, t_p) * \eta_p(y).$$

Then
$$\lambda(x, y) \leqslant \alpha(x, y) \leqslant \mu(x, y).$$

The sets $\alpha(x, t_p) * \eta_p(y)$ $(p = 0, 1, 2, \ldots, n-1)$ are disjoint, and by Theorem 10.8.1 on product sets,

$$m\{\alpha(x, t_p) * \eta_p(y)\} = m\{\alpha(x, t_p)\}m\{\eta_p(y)\}$$
$$= m(t_p, f)(t_{p+1} - t_p),$$

where $m(t, f)$ is the measure of the set $\{x; f(x) > t\}$.

Hence the measure of the set of points with indicator $\mu(x, y)$ is

$$m(\mu) = \sum_{p=0}^{n-1} m(t_p, f)(t_{p+1} - t_p).$$

Similarly
$$m(\lambda) = \sum_{p=0}^{n-1} m(t_{p-1}, f)(t_{p+1} - t_p),$$

and
$$m(\mu) - m(\lambda) = \sum_{p=0}^{n-1} \{m(t_p, f) - m(t_{p+1}, f)\}(t_{p+1} - t_p)$$
$$\leqslant \epsilon \sum_{p=0}^{n-1} \{m(t_p, f) - m(t_{p+1}, f)\} \leqslant \epsilon m(E),$$

if $\epsilon = \max(t_{p+1} - t_p)$ for $p = 0, 1, 2, ..., n-1$.

Therefore
$$m(\lambda) \leqslant \int_0^B m(t, f) \, dt \leqslant m(\mu)$$

and, since $\alpha \leqslant \mu$, $m^*(\alpha) \leqslant m(\mu)$.

If ω is an interval covering α, then
$$\omega - \alpha \leqslant \omega - \lambda$$

and, by the preceding result,
$$m^*(\omega - \alpha) \leqslant m(\omega - \lambda) = m(\omega) - m(\lambda),$$

whence
$$m(\lambda) \leqslant m(\omega) - m^*(\omega - \alpha) = m_*(\alpha).$$

Now
$$m^*(\alpha) - m_*(\alpha) \leqslant m(\mu) - m(\lambda) \leqslant \epsilon m(E).$$

Since this holds for each $\epsilon > 0$,
$$m^*(\alpha) = m_*(\alpha).$$

Thus the ordinate set α is measurable.

Hence
$$m(\lambda) \leqslant m(\alpha) \leqslant m(\mu)$$

and therefore
$$m(\alpha) = \int_0^B m(t, f) \, dt.$$

If the Lebesgue integral is *defined* geometrically as in Theorem 10.8.1, then it is possible to give very compact proofs of the convergence theorems (8.7.7 and 9.3.7) and of Fubini's theorem, as in Burkill's monograph (1953).

10.10. Fubini's theorem in *d* dimensions

The generalization of Fubini's theorem to d dimensions is proved in exactly the same way as in two dimensions and it is sufficient to state the results without proof if we use the customary compact notation.

The d-dimensional space R is the Cartesian product of the p-dimensional space $R^{(p)}$ and the q-dimensional space $R^{(q)}$ ($d = p+q$). The vector $x = (x_1, x_2,..., x_p)$ is a point in $R^{(p)}$ and the vector $y = (y_1, y_2,..., y_q)$ is a point in $R^{(q)}$. The indicator of a set of points E in R is a function

$$\alpha(x,y) = \alpha(x_1, x_2,..., x_p,\ y_1, y_2,..., y_q).$$

If E has finite d-dimensional measure in R,

$$\mu = \int_R \alpha(x,y),$$

then the set of points $\alpha(s,y)$ (s a fixed vector in $R^{(p)}$, $y \in R^{(q)}$) is measurable for almost all s and its q-dimensional measure is

$$\mu(s) = \int_{R^{(q)}} \alpha(s,y).$$

Also $\mu(s)$ is a measurable function of s if $s \in R^{(p)}$ and

$$\int_{R^{(q)}} \mu(s)\ \mathrm{d}s = \mu.$$

If $f(x,y) = f(x_1, x_2,..., x_p,\ y_1, y_2,..., y_q)$ is non-negative and summable over R then $f(x,t)$ is also summable over $R^{(p)}$, t being a fixed vector in $R^{(q)}$.

If
$$g(y) = \int_{R^{(p)}} f(x,y),$$

then $g(y)$ is summable over $R^{(q)}$ and

$$\int_{R^{(q)}} g(y) = \int_R f(x,y).$$

Finally, if $f(x,y)$ is non-negative and measurable in R, and if one repeated integral

$$\int_{R^{(q)}} \int_{R^{(p)}} f(x,y)$$

exists, then so also does the other

$$\int_{R^{(p)}} \int_{R^{(q)}} f(x,y)$$

and the two repeated integrals are equal.

10.11. Exercises

1. $f(x,\lambda)$ possesses a partial derivative

$$g(x,\lambda) = \partial f(x,\lambda)/\partial\lambda$$

if $a \leqslant x \leqslant b$, and $\alpha \leqslant \lambda \leqslant \beta$. $g(x,\lambda)$ is a bounded and measurable function of (x,λ). Prove that

$$h(\lambda) = \int\limits_a^b g(x,\lambda)\,\mathrm{d}x$$

exists and that $\displaystyle\int\limits_\alpha^\lambda h(t)\,\mathrm{d}t = \int\limits_a^b \{f(x,\lambda) - f(x,\alpha)\}\,\mathrm{d}x.$

Deduce that $\displaystyle h(\lambda) = \frac{\partial}{\partial\lambda}\int\limits_a^b f(x,\lambda)\,\mathrm{d}x$ p.p. in λ.

2. Use Fubini's theorem to prove that

$$\int\limits_0^\infty e^{-x^2}\,\mathrm{d}x \int\limits_0^\infty e^{-y^2}\,\mathrm{d}y = \tfrac{1}{4}\pi,$$

expressing the multiple integral as a Young integral.

3. If $g(x)$ and $h(x)$ are summable functions and

$$G(x) = \int\limits_{-\infty}^x g(s)\,\mathrm{d}s, \qquad H(x) = \int\limits_{-\infty}^x h(s)\,\mathrm{d}s,$$

prove that

$$\int\limits_{-\infty}^\infty g(x)H(x)\,\mathrm{d}x + \int\limits_{-\infty}^\infty h(x)G(x)\,\mathrm{d}x = \lim_{t\to\infty} G(t)H(t).$$

4. If $f(x,y)$ is bounded and measurable in the interval $(a \leqslant x \leqslant b,\ p \leqslant y \leqslant q)$ and is non-increasing in y for each fixed value of x, prove directly that

$$\int\limits_a^b \mathrm{d}x \int\limits_p^q f(x,y)\,\mathrm{d}y = \int\limits_p^q \mathrm{d}y \int\limits_a^b f(x,y)\,\mathrm{d}x.$$

5. If $f(x)$ is non-negative, and bounded over E, and if the ordinate set of $f(x)$ over E is measurable with measure m, show that $f(x)$ is measurable over E and that $m = \int_E f(x)\,\mathrm{d}x$ (Williamson, pp. 54, 55).

6. Prove Theorems 10.8.1 and 10.9.1 for unbounded measurable sets E.

11 The Lebesgue–Stieltjes integral

11.1. Introduction

The concept of the Stieltjes integral can be illustrated by the problem of calculating the quantity of heat Q required to raise the temperature of a given heterogeneous body by 1 °C, given the specific heat at each point of the body (Lebesgue 1928, p. xii).

Let the body be divided into a finite number of parts of *masses* $m_1, m_2,..., m_n$, and let $\underline{c}_p$ and $\bar{c}_p$ be the infimum and the supremum of the specific heat at points in the part with mass m_p. Then, by the definition of specific heat, Q is intermediate in value between the sums

$$\Lambda = \sum_{p=1}^{n} \underline{c}_p\, m^p \quad \text{and} \quad \mathrm{M} = \sum_{p=1}^{n} \bar{c}_p\, m_p.$$

As in § 3.3 we can consider the collections of numbers Λ and M for all subdivisions of the body into a finite number of parts, and we can define sup Λ and inf M as the lower and upper Stieltjes integrals of the specific heat over the mass of the body. If these bounds are equal we can define their common value to be the Stieltjes integral of the specific heat over the mass of the body.

This process is clearly analogous to that by which we obtained the lower and upper Darboux integrals (§ 3.3) and the Riemann integral, and the result is often called the Riemann–Stieltjes integral. It is subject to the same criticism as the Riemann integral and we shall therefore pass on at once and construct the analogue of the Lebesgue integral and thus obtain the Lebesgue–Stieltjes integral.

For this purpose we can adopt the whole of the Lebesgue theory if we make one small but vital change at the very beginning and replace the geometric measure of an interval by what we shall call the 'weighted measure'. Thus in the example quoted at the beginning of this section the primary concept would be not the *volumes* of the parts into which the body is divided but the *masses* of those parts. Now the physical concept of mass has one important property that is analogous to the mathematical concept of measure, viz. it is an *additive* function of the parts into which a body is divided, i.e. if a body of mass m is divided into a finite number of parts with masses $m_1, m_2,..., m_n$, then

$$m = m_1+m_2+...+m_n.$$

But the concept of mass differs from the concept of measure in as much as we can have masses that are concentrated into surfaces, lines, or points, whereas the three-dimensional measure of a plane, or straight line, or a point is zero. (We do not speak of three-dimensional measure of a surface or a curve in general, because there are pathological examples that invalidate the corresponding plausible assertion.) This 'grittiness' or lack of smoothness in a mass distribution necessitates a rather careful definition of the concept of weighted measure.

In three-dimensional space and even in two-dimensional space this leads to rather tiresome complications and we shall therefore first restrict ourselves to the Lebesgue–Stieltjes integral in one dimension.

11.2. The weighted measure

Whether we are considering an open interval (a, x) or a closed interval $[a, x]$ its weighted measure $w(x)$ is a non-negative, monotone, non-decreasing function of x. In any interval $a \leqslant x \leqslant b$, such a function is necessarily continuous at each point with the possible exception of a finite or enumerable set of points $\{x_n\}$ ($n = 1, 2,...$). At each of these points there exist the limits

$$w(x_n-0) = \lim w(x_n-h)$$

and

$$w(x_n+0) = \lim w(x_n+h)$$

as h tends to zero through positive values. Hence we are led to the following definitions.

DEFINITION 11.2.1. The weight function $w(x)$ is a non-negative, non-decreasing function of x.

DEFINITION 11.2.2 The weighted measure of a point ξ is
$$w(\xi+0)-w(\xi-0).$$
(This is zero unless ξ is one of the points of discontinuity $x_1, x_2,\dots .$)

The weighted measure of an open interval (a,b) is
$$w(b-0)-w(a+0).$$

The weighted measure of a closed interval $[a,b]$ is
$$w(b+0)-w(a-0).$$

The weighted measure of a half-open interval $[a,b)$ is
$$w(b-0)-w(a-0).$$

The theory of weighted measure can now be developed by exact analogy with the theory of geometric measure (Chapter 6).

DEFINITION 11.2.3. The weighted measure $w(\alpha)$ of an elementary set α, consisting of a finite number of disjoint intervals $\alpha_1, \alpha_2,\dots, \alpha_n$ is the sum
$$w(\alpha) = w(\alpha_1)+w(\alpha_2)+\dots+w(\alpha_n).$$

DEFINITION 11.2.4. The weighted measure $w(\alpha)$ of an outer set α consisting of an enumerable collection of disjoint intervals is the sum
$$w(\alpha) = \sum_{n=1}^{\infty} w(\alpha_n).$$

It may now be verified as in Theorem 6.3.6 for Lebesgue measure that, with these definitions, if σ and τ are bounded outer sets with the representations
$$\sigma = \sum_{s=1}^{\infty} \alpha_s \quad \text{and} \quad \tau = \sum_{t=1}^{\infty} \beta_t,$$
and if σ is covered by τ, then $\sum_{s=1}^{\infty} w(\alpha_s) \leqslant \sum_{t=1}^{\infty} w(\beta_t)$.

DEFINITION 11.2.5. The outer weighted measure $w^*(\sigma)$ of a bounded set of points σ is the infimum of the weighted measures $w(\alpha)$ of the outer sets α which cover σ, i.e.
$$w^*(\sigma) = \inf w(\alpha) \quad \text{for} \quad \alpha \geqslant \sigma.$$

DEFINITION 11.2.6. The inner weighted measure $w_\omega(\sigma)$ of a bounded set of points σ with respect to an interval ω which covers σ is
$$w_\omega(\sigma) = w(\omega) - w^*(\omega - \sigma).$$

THEOREM 11.2.1. *The value of $w_\omega(\sigma)$ is independent of ω.*

DEFINITION 11.2.7. The inner weighted measure $w_*(\sigma)$ of a bounded set of points is the value of $w_\omega(\sigma)$ for any interval ω which covers σ.

DEFINITION 11.2.8. A bounded set of points σ is said to have Stieltjes measure $\mu_w(\sigma)$ with respect to the weight function $w(x)$ if the outer and inner weighted measures of σ are equal, and the value of the Stieltjes measure is defined to be
$$\mu_w(\sigma) = w^*(\sigma) = w_*(\sigma).$$

For brevity we often say that, under these conditions, 'the set σ is measurable (w)'.

THEOREM 11.2.2. *The Stieltjes measure $\mu_w(\sigma)$ is a positive, additive, continuous functional of the indicator $\sigma(x)$, i.e.*
$$\mu_w(\sigma) \geqslant 0,$$
$$\mu_w(\sigma_1 + \sigma_2) = \mu_w(\sigma_1) + \mu_w(\sigma_2) \quad if \ \sigma_1 \sigma_2 = 0,$$
and
$$\mu_w(\sigma) = \sum_{n=1}^{\infty} \mu_w(\sigma_n)$$
if $\sigma_1, \sigma_2,\dots$ is an enumerable collection of disjoint sets each bounded by the same interval I.

The Stieltjes measure $\mu_w(\sigma)$ therefore possesses many of the properties of an integral, and is therefore commonly written in the form
$$\mu_w(\sigma) = \int \sigma(x) \, dw(x).$$

In particular, for an interval I,
$$\mu_w(I) = \int_I dw(x).$$

DEFINITION 11.2.9. A bounded function $f(x)$ will be said to have Stieltjes measure function $\mu_w(t, f)$ (or to be measurable w) if the set of points $E\{x, f(x) > t\}$ has Stieltjes measure $\mu_w(t, f)$ for each value of t.

As in § 8.5 we can introduce the upper and lower Lebesgue bracketing functions $\mu(x)$ and $\lambda(x)$, corresponding to a partition

$$A = t_0 < t_1 < t_2 < ... < t_n = B$$

of the range of $f(x)$ for $a \leqslant x \leqslant b$. Let

$$\epsilon = \max(t_{p+1}-t_p) \quad \text{for } p = 0, 1, 2,..., n-1.$$

THEOREM 11.2.3.

$$\int_a^b \mu(x)\,\mathrm{d}w(x) - \int_a^b \lambda(x)\,\mathrm{d}w(x) \leqslant \epsilon\{\mu_w(B,f)-\mu_w(A,f)\}.$$

DEFINITION 11.2.10. The Lebesgue–Stieltjes integral of $f(x)$ with respect to the weight function $w(x)$ over the interval $[a,b]$ is

$$\int_a^b f(x)\,\mathrm{d}w(x) = \inf \int_a^b \mu(x)\,\mathrm{d}w(x) = \sup \int_a^b \lambda(x)\,\mathrm{d}w(x),$$

for all Lebesgue bracketing functions $\lambda(x)$ and $\mu(x)$.

THEOREM 11.2.4.

$$\int_a^b f(x)\,\mathrm{d}w(x) = \mu_w(A,f).A + \int_A^B \mu_w(t,f)\,\mathrm{d}t.$$

This identifies the Lebesgue–Stieltjes integral with the 'Young–Stieltjes' integral of the monotone measure function $\mu_w(t,f)$.

We can then extend the definition to unbounded functions and unbounded intervals of integration as before, by considering separately the positive and negative parts of $f(x)$.

11.3. The Lebesgue representation of a Stieltjes integral

Lebesgue has shown that the Lebesgue–Stieltjes integral in one dimension can be represented as an ordinary Lebesgue integral by a simple transformation of the independent variable from x to the 'Lebesgue inverse function' of the weight function $w(x)$.

The weight function $w(x)$ has no unique inverse in the ordinary sense, for, to a prescribed value of y, there may correspond a whole interval, $\alpha \leqslant x \leqslant \beta$ of points at which $w(x) = y$ as in Fig. 3 (a). However, since $w(x)$ is non-decreasing in any bounded interval $[a, b]$, it is measurable in the sense of Lebesgue, and its

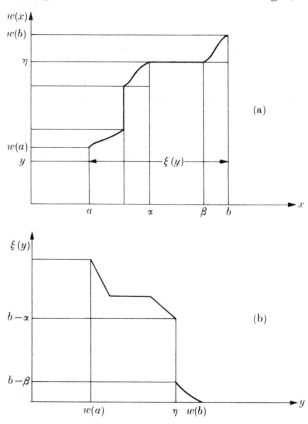

Fig 3.

measure function $\xi(y)$, the Lebesgue measure of the set of points $E\{x;\, w(x) > y\}$, does provide a species of inverse function for $w(x)$ (see Fig. 3 (b)).

The Lebesgue inverse function $\xi(y)$ is a non-increasing function which is discontinuous at each value η of y which corresponds to an interval $[\alpha, \beta]$ of x in which the weight function $w(x)$ remains constant, for

$$\xi(\eta+0) = b-\beta \quad \text{and} \quad \xi(\eta-0) = b-\alpha.$$

In the Lebesgue transformation we introduce the function defined by the relation

$$f[\xi(y)] = \phi(y).$$

The indicator function of $\phi(y)$, say $\beta(y, t)$ is defined by the relations

$$\beta(y, t) = \begin{cases} 1 & (\phi(y) > t), \\ 0 & (\phi(y) \leqslant t). \end{cases}$$

If $\alpha(x, t)$ is the indicator of $f(x)$, then these relations imply that

$$\alpha(\xi(y), t) = \beta(y, t).$$

Hence the lower Lebesgue bracketing function for $f(x)$ is

$$\lambda = \sum_{p=0}^{n-1} t_p \{\alpha(x, t_p) - \alpha(x, t_{p+1})\}$$

$$= \sum_{p=0}^{n-1} t_p \{\beta(y, t_p) - \beta(y, t_{p+1})\}$$

and its Stieltjes integral is

$$\int_a^b \lambda(x)\, dw(x) = \sum_{p=0}^{n-1} t_p \{m(t_p) - m(t_{p+1})\},$$

where

$$m(t_p) = \int \beta(y, t_p)\, dy,$$

which is the Lebesgue measure of the set $E\{y; \phi(y) > t_p\}$.

Therefore, the supremum of $\int_a^b \lambda(x)\, dw(x)$ is the Lebesgue integral of $\phi(y)$ over the range $y = w(a)$ to $y = w(b)$. We could similarly identify the infimum of $\int_a^b \mu(x)\, dw(x)$. But the common value of these bounds is the Lebesgue–Stieltjes integral of $f(x)$ with respect to $w(x)$.

Therefore

$$\int_a^b f(x)\, dw(x) = \int_{w(a)}^{w(b)} f[\xi(y)]\, dy.$$

In particular, if $f(x) = 1$ in the interval (a, s) and is zero elsewhere, then

$$\int_a^s f(x)\, dw(x) = \int_{w(a)}^{w(s)} 1\, dy = w(s) - w(a).$$

11.4. The Lebesgue–Stieltjes integral in one dimension

The preceding section has been devoted to the Lebesgue–Stieltjes integral of a bounded function $f(x)$ with respect to a non-negative, non-decreasing weight function $w(x)$. The theory is easily extended to weight functions that are of bounded variation, and we shall merely state the relevant definitions and theorems, leaving the proofs to the reader.

THEOREM 11.4.1. *If $w_1(x)$ and $w_2(x)$ are two non-negative, non-decreasing weight functions, so also is their sum*

$$w(x) = w_1(x) + w_2(x);$$

and if a bounded function $f(x)$ has Stieltjes measure for an interval I with respect to each of the weight functions $w_1(x)$ and $w_2(x)$, then it has Stieltjes measure with respect to $w(x)$ and

$$\int_I f(x)\,\mathrm{d}w(x) = \int_I f(x)\,\mathrm{d}w_1(x) + \int_I f(x)\,\mathrm{d}w_2(x).$$

DEFINITION 11.4.1. A function $w(x)$ of the real variable x is said to be 'of bounded variation' in an interval I if there exists a pair of non-negative, non-decreasing functions, $p(x)$ and $n(x)$, such that
$$w(x) = p(x) - n(x) \quad \text{for } x \in I.$$

THEOREM 11.4.2. *If $w(x)$ has bounded variation in I and $[p_1(x), n_1(x)]$, $[p_2(x), n_2(x)]$ are two pairs of non-negative, non-decreasing functions such that*

$$w(x) = p_1(x) - n_1(x) = p_2(x) - n_2(x),$$

and if the bounded function $f(x)$ has Stieltjes measure for an interval I with respect to each of the weight functions $p_1(x)$, $p_2(x)$, $n_1(x)$, $n_2(x)$, then

$$\int_I f(x)\,\mathrm{d}p_1(x) - \int_I f(x)\,\mathrm{d}n_1(x) = \int_I f(x)\,\mathrm{d}p_2(x) - \int_I f(x)\,\mathrm{d}n_2(x).$$

DEFINITION 11.4.2. The Lebesgue–Stieltjes integral of the bounded function $f(x)$ with respect to the weight function $w(x)$ of Definition 11.4.1 for an interval I is

$$\int_I f(x)\, \mathrm{d}w(x) = \int_I f(x)\, \mathrm{d}p_1(x) - \int_I f(x)\, \mathrm{d}n_1(x)$$

in the notation of Theorem 11.4.2.

11.5. The Lebesgue–Stieltjes integral in two dimensions

The theory of the Lebesgue–Stieltjes integral in two or more dimensions will be found, in summary form, in modern books on statistics and the theory of probability (e.g. Moran 1969, p. 203 and Kingman and Taylor 1966, p. 95) and, in more detail, in serious works on integration (e.g. McShane 1947, chap. vii). The crux of the theory is the introduction of a weight function $w(I)$ which is a mapping of the intervals in Euclidean space of d dimensions into the real axis, and which is

(i) non-negative;

(ii) additive, in the sense that if the interval I is the union of two disjoint intervals I_1, I_2 then

$$w(I) = w(I_1) + w(I_2).$$

The main difficulty arises from the possible discontinuities in the weight function.

From the point of view of the physicist the weight function may represent a discrete distribution of mass at a number of isolated points. The question then arises, does the weight function $w(I)$ include any point masses on the boundaries of I? And, if so, how is the additive character of $w(I)$ preserved?

From the point of view of the mathematician the possibility of defining the weight function for any bounded open set Ω depends upon the representation of Ω as the union of an enumerable collection of disjoint intervals I_n, and hence the intervals I_n must not all be open intervals, or all closed intervals.

After meditating on these difficulties the reader may be prepared to settle for the following definitions, which allow a fairly concise account of the theory. In order to exhibit the essence of the theory we shall restrict ourselves to two dimensions. The extension to three or more dimensions is an obvious generalization, which is fully discussed in the references given above.

DEFINITION 11.5.1. In the convenient terminology of Ingleton (1965) a 'standard' interval is the set of points (x, y) in the half-open rectangle

$$a \leqslant x < b, \qquad c \leqslant y < d.$$

THEOREM 11.5.1. *An open set of points is the union of an enumerable collection of disjoint standard intervals.*

DEFINITION 11.5.2. A weight function $w(x,y)$ is a non-negative function of the real variables x, y, which is non-decreasing in x, non-decreasing in y, and which is semi-continuous in the sense that

$$w(x, y) = \lim w(x-\epsilon, y)$$
$$= \lim w(x, y-\epsilon)$$

as ϵ tends to zero through positive values.

DEFINITION 11.5.3. A Stieltjes measure function $w(I)$ of the standard intervals $I: a \leqslant x < b, \ c \leqslant y < d$, is a function of the form
$$w(a, c) - w(b, c) + w(b, d) - w(a, d),$$
where $w(x, y)$ is a weight function according to Definition 11.5.2, and where $w(I) \geqslant 0$ if $a < b, \ c < d$.

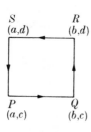

S (a,d) R (b,d)

P (a,c) Q (b,c)

FIG. 4

Thus from the point of view of the physicist the measure function $w(I)$ includes any concentrated masses on the edges PQ and SP but excludes any concentrated masses on the edges QR and RS (Fig. 4).

THEOREM 11.5.2. *The Stieltjes measure function $w(I)$ of Definition 11.5.2 is non-negative and additive for all standard intervals I.*

We can now follow the same process as in Chapter 7 and proceed to define the outer and inner Stieltjes measures of a set of points Ω by covering Ω with an enumerable collection of disjoint standard intervals. We can then define the sets of points that have a unique Stieltjes measure.

A bounded function $f(x, y)$ will be said to have Stieltjes measure if this is true of each set of points (x, y) for which $f(x, y) > t$.

We can then introduce the indicator $\alpha(x, y, t)$ of this set of points and its Stieltjes measure $\mu_w(t)$ which we identify with the Lebesgue–Stieltjes integral of the indicator.

We can bracket $f(x, y)$ by two simple functions as in § 11.2 and define the Lebesgue–Stieltjes integral of $f(x)$ over a region Ω with respect to a weight function $w(x, y)$.

As in one dimension we can consider weight functions of bounded variation and functions that are summable with respect to such weight functions.

11.6. Exercises

1. The Heaviside unit function

$$H(x) = \begin{cases} 0 & (x \leqslant 0), \\ 1 & (x > 0), \end{cases}$$

is a weight function in one dimension, which is semi-continuous on the right, i.e. $H(x) = \lim H(x - \epsilon)$, as ϵ tends to zero through positive values. It represents a unit mass at the origin.

2. Construct the corresponding weight function in two dimensions for a unit mass at the origin.

3. In d dimensions, where $d \geqslant 2$, a standard interval is specified by the 2^d vertices, $V_1, V_2, \dots$ of a rectangular block I,

$$a_k \leqslant x_k < b_k \quad (k = 1, 2, \dots, d).$$

The numbers $a_1, a_2, \dots, a_d$ will be called the lower coordinates of I, and the numbers $b_1, b_2, \dots, b_d$ will be called the upper coordinates of I. If a vertex V_i is specified by r upper coordinates and by $d - r$ lower coordinates, we assign to it the number

$$N(V_i) = (-1)^{r+d}.$$

Show that a Stieltjes measure function in d dimensions has the form

$$\mu(I) = \sum N(V_i) f(V_i) \quad (1 \leqslant i \leqslant 2^d),$$

where $f(V_i)$ is a non-negative function of the coordinates of V_i, non-decreasing in each coordinate and semi-continuous in the sense that

$$f(x_1, x_2, \dots, x_d) = \lim f(x_1 - \epsilon_1, x_2 - \epsilon_2, \dots, x_d - \epsilon_d)$$

as $\epsilon_1, \epsilon_2, \dots, \epsilon_d$ tend to zero through positive values.

12 Epilogue

12.1. The generality of the Lebesgue integral

It seems appropriate to conclude this introductory account of the Lebesgue integral with a rapid summary of its advantages.

The most obvious advantage of the Lebesgue theory is in its application to the integration of sequences of functions where the generality and simplicity of the theorems of bounded and dominated convergence greatly facilitate the analysis. By comparison the Riemann theory is much restricted in scope.

But a greater attraction of the Lebesgue theory is its wide generality and the kind of inner necessity and inevitability with which the theory develops. From this point of view the central feature is the relation between an integrable function $f(x)$ and its indicator $\alpha(x,t,f)$.

To exhibit this characteristic of the Lebesgue theory we shall develop the descriptive definition of the Lebesgue integral as given by Lebesgue himself in the second edition (1928) of his *Leçons sur l'intégration*.

12.2. The descriptive definition of the Lebesgue integral

To avoid confusion with the Lebesgue integral as constructively defined in the preceding chapters we shall now speak of the 'Lebesgue functional'.

DEFINITION 12.2.1. A Lebesgue functional is a functional $\int_a^b f(x)\,dx$, defined in a space $\mathscr{L}$ of bounded and real-valued functions $f(x)$ on the interval (a,b), which is linear (L), positive (P), 'absolute' (A), and 'monotonely convergent' (C), with the Lebesgue normalization (N).

The functions in $\mathscr{L}$ will be called 'Lebesgue functions'.

The definitions of linear, positive functionals have been given in § 2.2. Condition (A) implies that, if $f(x) \in \mathscr{L}$, so also does its absolute value $|f(x)|$, and condition (N) implies that the indicator $\alpha(x)$ of any interval (p, q) is a Lebesgue function, and that

$$\int_a^b \alpha(x)\, dx = q - p \quad \text{if } a \leqslant p < q \leqslant b.$$

Condition (C) means that, if $\{f_n(x)\}$ is a sequence of Lebesgue functions, such that

$$0 \leqslant f_n(x) \leqslant f_{n+1}(x) \leqslant M$$

for all n and $a \leqslant x \leqslant b$, where M is independent of x and n, and if the sequence $\{f_n(x)\}$ converges to a limit function $f(x)$ at all points of (a, b) except at an enumerable set $x = x_1, x_2,...,$ then $f(x)$ is a Lebesgue function in (a, b) and

$$\lim_{n \to \infty} \int_a^b f_n(x)\, dx = \int_a^b f(x)\, dx.$$

Condition (L) now implies that the space $\mathscr{L}$ contains the step functions of § 2.3.

It is, of course, necessary to verify that these characteristics by which we have described the Lebesgue functional are mutually consistent. To do this it is sufficient to observe that all these characteristics are possessed by the integrals of functions which are constant in the interval (a, b).

THEOREM 12.2.1. *If $f(x) = c$ in (a, b), the functional*

$$\int_a^b f(x)\, dx = c(b-a)$$

possesses the characteristics (L), (P), (A), (C), *and* (N).

For the characteristics (L), (P), (A), and (N) the proof is too elementary to print. For the characteristic (C), let

$$f_n(x) = c_n \quad \text{in } (a, b),$$

$$0 \leqslant c_n \leqslant c_{n+1} \leqslant M,$$

and

$$c_n \to c \quad \text{as } n \to \infty.$$

Then
$$\int_a^b f_n(x)\,\mathrm{d}x = c_n(b-a) \to c(b-a) = \int_a^b f(x)\,\mathrm{d}x,$$
as $n \to \infty$.

In order to show that the class $\mathscr{L}$ of Lebesgue functions does contain some non-trivial functions we give the following theorem.

THEOREM 12.2.2. *Any bounded, non-negative, non-decreasing function $f(x)$ is a Lebesgue function.*

Let
$$f(a) = A, \qquad f(b) = B,$$
and
$$b-a = 2^n \epsilon_n,$$
where n is a positive integer. Define the step function, $\phi_n(x)$, by the condition that
$$\phi_n(x) = f(a+p\epsilon_n),$$
if
$$a+p\epsilon_n \leqslant x < a+(p+1)\epsilon_n,$$
for $p = 0, 1, 2,..., 2^n-1$. Then, if x is fixed, p is the greatest integer not exceeding
$$2^n(x-a)/(b-a).$$
Similarly
$$\phi_{n+1}(x) = f(a+q\epsilon_{n+1})$$
if
$$a+q\epsilon_{n+1} \leqslant x < a+(q+1)\epsilon_{n+1},$$
for $q = 0, 1, 2,..., 2^{n+1}-1$. If x is fixed, q is the greatest integer not exceeding $2^{n+1}(x-a)/(b-a)$, but
$$2p \leqslant 2^{n+1}(x-a)/(b-a),$$
whence
$$2p \leqslant q.$$
Therefore
$$(a+q\epsilon_{n+1})-(a+p\epsilon_n) = 2^{-n-1}(b-a)(q-2p) \geqslant 0.$$
But $f(x)$ is non-decreasing, whence
$$\phi_{n+1}(x)-\phi_n(x) = f(a+q\epsilon_{n+1})-f(a+p\epsilon_n) \geqslant 0.$$
Also
$$0 \leqslant x-(a+p\epsilon_n) < \epsilon_n.$$
Hence at each value of x, whereat $f(x)$ is continuous,
$$\phi_n(x) \to f(x) \quad \text{as } n \to \infty.$$
But $f(x)$ is non-decreasing, and hence is discontinuous only at an enumerable set of points in (a,b). Therefore by condition (C), $f(x)$ is a Lebesgue function.

12.3. Measure functions

DEFINITION 12.3.1. The indicator $\alpha(x,t,f)$ of a function $f(x)$ is defined by the relations

$$\alpha(x,t,f) = \begin{cases} 1 & (f(x) > t), \\ 0 & (f(x) \leqslant t), \end{cases}$$

where t is any real number.

THEOREM 12.3.1. *If $f(x)$ is a Lebesgue function in (a,b), so also is its indicator $\alpha(x,t,f)$.*

Let $\alpha_n(x) = \frac{1}{2}|nf(x)-nt| - \frac{1}{2}|nf(x)-nt-1| + \frac{1}{2}$,

where n is a positive integer. Then it follows from conditions (N), (L), and (A) of Definition 12.2.1, that $\alpha_n(x)$ is a Lebesgue function in (a,b). Now

$$\alpha_n(x) = \begin{cases} 1 & (f(x) \geqslant t+1/n), \\ nf(x)-nt & (t \leqslant f(x) \leqslant t+1/n), \\ 0 & (f(x) \leqslant t). \end{cases}$$

Hence $\{\alpha_n(x)\}$ is a bounded, non-decreasing sequence of Lebesgue functions.

Also $\alpha_n(x) \to \begin{cases} 1 & (f(x) > t), \\ 0 & (f(x) \leqslant t), \end{cases}$

as $n \to \infty$. Therefore, by condition (C) of Definition 12.2.1, the indicator $\alpha(x,t,f)$ is a Lebesgue function in (a,b).

DEFINITION 12.3.2. The Lebesgue integral,

$$m(t,f) = \int_a^b \alpha(x,t,f)\, dx,$$

which exists by Theorem 12.3.1, is called the 'measure function' of $f(x)$.

THEOREM 12.3.2. *If $f(x)$ is a Lebesgue function in (a,b), and if, in this interval,* $A < f(x) < B$,

then the measure function $m(t,f)$ is a Lebesgue function of t in the interval $A < t < B$.

M

LEMMA. *If $\phi(x)$ and $\psi(x)$ are Lebesgue functions in (a,b) and if $\phi(x) \geqslant \psi(x)$ in (a,b), then*

$$\int_a^b \phi(x)\,dx \geqslant \int_a^b \psi(x)\,dx.$$

For by conditions (L) and (P) of Definition 12.2.1,

$$\int_a^b \phi(x)\,dx - \int_a^b \psi(x)\,dx = \int_a^b \{\phi(x)-\psi(x)\}\,dx.$$

Now, if $s < t$, then

$$\alpha(x,s,f) \geqslant \alpha(x,t,f),$$

whence $m(s,f) \geqslant m(t,f)$, by the lemma. Thus $-m(t,f)$ is a non-decreasing function of t.

Also $\qquad\qquad \alpha(x,A,f) = 1$

and $\qquad\qquad \alpha(x,B,f) = 0, \quad$ if $a < x < b$.

Therefore $\qquad m(A,f) = b-a \quad$ by (N)

and $\qquad m(B,f) = \int_a^b 0\,dx$

$$= \int_a^b 1\,dx - \int_a^b 1\,dx \quad \text{by (L)}$$

$$= 0 \quad \text{by (N)}.$$

Thus $\qquad 0 = m(B,f) \leqslant m(t,f) \leqslant m(A,f),$

if $A < t < B$. Hence the function $-m(t,f)$ is bounded. Therefore the function $-m(t,f)$, and also (by condition (L)), the function $m(t,f)$ are Lebesgue functions of t in (A, B).

12.4. The Young integral

THEOREM 12.4.1. *If $f(x)$ is a Lebesgue function in (a,b), and if*

$$0 \leqslant f(x) \leqslant M$$

in this interval, then

$$\int_a^b f(x)\,dx = \int_0^M m(t,f)\,dt.$$

Let $0 = t_0 < t_1 < t_2 < ... < t_n = M$ be a partition of the range $[0, M]$ of $f(x)$.

As in § 8.5 we introduce the Lebesgue bracketing functions $\lambda(x)$ and $\mu(x)$:

$$\lambda(x) = \sum_{p=0}^{n-1} t_p\{\alpha(x,t_p,f) - \alpha(x,t_{p+1},f)\},$$

$$\mu(x) = \sum_{p=0}^{n-1} t_{p+1}\{\alpha(x,t_p,f) - \alpha(x,t_{p+1},f)\}.$$

Then $$\lambda(x) \leqslant f(x) \leqslant \mu(x).$$

Now by conditions (L), (N) and Theorem 12.3.1, $\lambda(x)$ and $\mu(x)$ are Lebesgue functions in (a,b). Hence, by the lemma of Theorem 12.3.2,

$$\int_a^b \lambda(x)\,\mathrm{d}x \leqslant \int_a^b f(x)\,\mathrm{d}x \leqslant \int_a^b \mu(x)\,\mathrm{d}x.$$

The analysis of § 8.6 also shows that

$$\int_a^b \lambda(x)\,\mathrm{d}x \leqslant \int_0^M m(t,f)\,\mathrm{d}t \leqslant \int_a^b \mu(x)\,\mathrm{d}x,$$

and that $$\int_a^b \mu(x)\,\mathrm{d}x - \int_a^b \lambda(x)\,\mathrm{d}x \leqslant \epsilon(b-a),$$

where $\epsilon = \max(t_{p+1}-t_p)$ for $p = 0, 1, 2,..., n-1$. Since this is true for any tolerance $\epsilon > 0$ it follows that

$$\int_a^b f(x)\,\mathrm{d}x = \int_0^M m(t,f)\,\mathrm{d}t.$$

This identifies the Lebesgue integral of $f(x)$ over (a,b) with the W. H. Young integral of its measure function $m(t,f)$ over the range $(0, M)$ of $f(x)$.

The significance of this investigation is that it proves that the domain of the Lebesgue functional, as defined in 12.2.1, is included in the space of bounded measurable functions, i.e. the functions $f(x)$ whose indicator $\alpha(x,t,f)$ is a Lebesgue function. Of course it remains to be proved that all such bounded measurable functions are in fact Lebesgue functions and, to do this, we need the constructive definitions of the preceding chapters.

The use of monotone sequences then allows us to extend the

definition of the Lebesgue integral to unbounded functions
and to unbounded intervals as in Chapter 9, but such integrals
are necessarily absolutely convergent.

If, however, we are prepared to waive condition (A) and admit
non-absolutely convergent integrals, then the space of integrable
functions can be considerably extended. In this work of analytic
exploration the initial advances were made by A. Denjoy
(1942–9) by means of constructive definitions involving trans-
finite induction.

Later, O. Perron gave a descriptive definition by adapting
the method of bracketing to give directly upper and lower
bounds to the integral (rather than the integrand, as in our
exposition). Outstanding advances in this field have been made
by A. J. Ward and R. Henstock and are described in the latter
author's book *Theory of integration* (1963).

The intentions of the author of the present work will be
amply fulfilled if this introductory account of the Lebesgue
integral has stimulated the reader to study the more formal and
profound account of the subject. The following works are
especially recommended for the undergraduate.

The original paper and books by Lebesgue and the exposition
by de la Vallée-Poussin are somewhat terse, and chapters x, xi,
and xii of *The theory of functions* by E. C. Titchmarsh (Clarendon
Press, Oxford) will be found to provide a most illuminating
commentary.

The *geometric* theory of the Lebesgue integral is expounded
with great clarity and conciseness in J. C. Burkill's Cambridge
Tract. A rather more advanced treatment is given with greater
emphasis on *topological* and *set-theoretic* concepts in *Lebesgue
integration* by J. H. Williamson. By contrast, the treatment by
A. N. Kolmogorov and S. V. Fomin in the translation published
by Academic Press (New York and London, 1961) with the
title *Measure, Lebesgue integrals and Hilbert space*, may be
perhaps described as more *algebraical* in presentation.

A forthcoming book by A. W. Ingleton will provide yet another
attractive line of approach to integration theory which may be
roughly characterized as the method of functional analysis.

12.5. References

BOAS, R. P. (1960). *A primer of real functions*. Wiley.

BURKILL, J. C. (1953). *The Lebesgue integral*. Cambridge University Press.

DENJOY, A. (1941–9). *Leçons sur le calcul des coefficients*. Gauthier-Villars, Paris.

HARTMAN, S., and MIKUSINSKI, J. (1961). *The theory of Lebesgue measure and integration*. Pergamon Press, Oxford.

HENSTOCK, R. (1963). *Theory of integration*. Butterworths, London.

HOBSON, E, W. (1927). *The theory of functions of a real variable and the theory of Fourier's series*, vols. i and ii. Cambridge University Press.

INGLETON, A. W. (1965). *Notes on integration*. The Mathematical Institute, Oxford.

KINGMAN, J. F. C., and TAYLOR, S. J. (1966). *Introduction to measure and probability*. Cambridge University Press.

KOLMOGOROV, A. N., and FOMIN, S. V. (1961). *Measure, Lebesgue integrals and Hilbert space*. Academic Press, New York.

LEBESGUE, H. (1902). Integrale, longueur, aire. *Annali Mat. pur. appl.* **3**, 231–359.

—— (1904, 1928). *Leçons sur l'intégration*. Gauthier-Villars, Paris.

MC SHANE, E. J. (1947). *Integration*. Princeton University Press.

MORAN, P. A. P. (1968). *An introduction to probability theory*. Clarendon Press, Oxford.

RIESZ, F., and SZ-NAGY, B. (1953). *Leçons d'analyse fonctionelle*. Akademiai Kiado, Budapest.

SAKS, S. (1937). *Theory of the integral* (English translation by L. C. Young). Subwencji Fundasza Kultury Narodowej, Warszawa–Lwów.

SOLOVAY R, M. (1971) *Annals of Math.* (2), **92**, 1.

DE LA VALLÉE-POUSSIN, C. J. (1915). *Trans. Am. math. Soc.* **16**, 435.

—— (1916). *Intégrales de Lebesgue*. Gauthier-Villars, Paris.

WILLIAMSON, J. H. (1962). *Lebesgue integration*. Holt, Rinehart, and Winston, New York.

YOUNG, L. C. (1927). *The theory of integration*. Cambridge University Press.

YOUNG, W. H. (1905). On upper and lower integrals. *Proc. Lond. math. Soc.* **2**, 52.

Index

PRINTED IN GREAT BRITAIN
AT THE UNIVERSITY PRESS, OXFORD
BY VIVIAN RIDLER
PRINTER TO THE UNIVERSITY